# Earth Building Practice

Dipl.-Ing. Ulrich Röhlen
Dr.-Ing. Christof Ziegert

# Earth Building Practice
## Planning – Design – Building

1st edition 2011

Beuth Verlag GmbH · Berlin · Vienna · Zurich

## Bauwerk

© 2011 Beuth Verlag GmbH
Berlin · Vienna · Zurich
Burggrafenstraße 6
10787 Berlin

Tel:      +49 30 2601-1
Fax:      +49 30 2601-1260
Internet: www.beuth.de
Email:    info@beuth.de

All rights reserved, whether the whole or part of the material is concerned, specifically the rights of translation, reprinting, re-use of illustrations, recitation, broadcasting, reproduction on microfilms or in other ways, and storage in data banks. For any kind of use, permission of the copyright owner must be obtained.

Neither the author nor the publisher make any warranty or representation, expressed or implied, with respect to the information contained in this publication, or assume any liability with respect to the use of, or damages resulting from, this information. Unless otherwise stated, all illustrations are by the authors. Numerical data is supplied without liability.

Cover photo:   Jahili Fort, Al Ain, Abu Dhabi, Torsten Seidel Fotografie
Design and layout:   Julian Reisenberger, Weimar
Translation into English:   Julian Reisenberger, Weimar
Printed and bound: fgb, freiburger graphische betriebe GmbH & Co. KG, Freiburg

Printed on acid-free and age-resistant paper in accordance with DIN EN ISO 9706.

ISBN 978-3-410-21737-4

# Table of contents

Foreword .................................................................................... 1

## 1 Earth building today .................................................. 3
1.1 **The state of the art of earth building materials** ........................... 3
1.2 **Reasons for using earth building materials** ................................ 4

## 2 The raw material – soils for construction ............ 7
2.1 **Introduction** ............................................................................. 7
2.2 **Types of soils and their formation** ............................................ 8
2.3 **Clay minerals – the binding agent** ............................................ 8
2.3.1 Structure and cohesion ............................................................ 8
2.3.2 Two layer clay minerals ........................................................... 11
2.3.3 Three layer clay minerals ......................................................... 12
2.4 **Testing the suitability of earth mixtures** ................................... 14
2.4.1 In-situ soil testing ................................................................... 14
2.4.2 Laboratory testing of earth mixtures ....................................... 16
2.4.2.1 Grain size distribution ............................................................ 16
2.4.2.2 Binding force test (figure-8-shape test) ................................... 17
2.4.2.3 Determining the compressive strength and shrinkage of earth mixtures ..... 19
2.4.2.4 Qualitative determination of the natural lime content ............. 19
2.4.2.5 Determining soil salinity ........................................................ 20
2.5 **Processing** .............................................................................. 21
2.5.1 Natural processing of earth mixtures ...................................... 22
2.5.2 Mechanical processing ........................................................... 22

| | | |
|---|---|---|
| **3** | **Earth building materials – composition and properties** | **25** |
| 3.1 | **Aggregates and additives** | 25 |
| 3.1.1 | Aggregates | 25 |
| 3.1.2 | Additives | 29 |
| 3.2 | **Mixing** | 31 |
| 3.3 | **Properties** | 32 |
| | | |
| **4** | **Earth plasters** | **37** |
| 4.1 | **The application of earth plasters** | 37 |
| 4.1.1 | Reasons for using earth plasters | 37 |
| 4.1.2 | Application areas of earth plasters | 38 |
| 4.1.3 | Aspects of long-term use | 39 |
| 4.2 | **Constitution of earth plaster mortars** | 40 |
| 4.2.1 | General composition of earth plaster mortars | 40 |
| 4.2.2 | Kinds of earth plaster mortars | 41 |
| 4.2.3 | Grades of earth plaster mortar | 42 |
| 4.3 | **Substrates for earth plasters** | 43 |
| 4.3.1 | Plaster substrates in general | 43 |
| 4.3.2 | Common plaster substrates | 45 |
| 4.3.3 | Primers and absorption sealers | 48 |
| 4.3.4 | Plaster laths | 49 |
| 4.3.5 | Special considerations for substrates for coloured plasters | 50 |
| 4.3.6 | Surface quality levels of plaster substrates for subsequent plastering | 51 |
| 4.4 | **Earth plaster systems** | 59 |
| 4.4.1 | Earth plaster systems in general | 59 |
| 4.4.2 | Shrinkage cracks in undercoat layers | 59 |
| 4.4.3 | Reinforcement fabric | 61 |
| 4.4.4 | Earth plaster on surfaces subject to thermal fluctuation | 63 |
| 4.5 | **Processing and application** | 63 |
| 4.5.1 | Mortar preparation | 63 |
| 4.5.2 | Mortar application | 67 |
| 4.5.3 | Corner profiles, plaster beads and junctions | 68 |
| 4.5.4 | Surface finishing | 70 |
| 4.5.5 | Special considerations for coloured earth plaster surfaces | 71 |
| 4.5.6 | Drying | 71 |

| | | |
|---|---|---|
| 4.5.7 | Crack repairs and additional surface finishing | 77 |
| **4.6** | **Paints and wall coverings for earth plasters** | **78** |
| 4.6.1 | Painting and surface stabilisation | 78 |
| 4.6.2 | Wallpaper | 79 |
| 4.6.3 | Lime skim coat plaster | 80 |
| 4.6.4 | Wall tiles on earth plasters | 80 |
| **4.7** | **Requirements of earth plasters** | **80** |
| 4.7.1 | Mechanical properties | 80 |
| 4.7.2 | Building biology requirements | 83 |
| 4.7.3 | Moisture sorption capacity requirement | 85 |
| 4.7.4 | Visual requirements | 85 |
| **4.8** | **Guaranteeing material properties** | **85** |
| 4.8.1 | Basic testing procedures and declarations | 85 |
| 4.8.2 | Supplementary testing procedures and declarations | 86 |
| 4.8.3 | Quality control | 87 |
| **4.9** | **Building material and building element properties** | **88** |
| 4.9.1 | Mechanical properties | 88 |
| 4.9.2 | Thermal insulation and protection against moisture | 88 |
| 4.9.3 | Sound insulation and acoustics | 89 |
| 4.9.4 | Fire performance | 90 |

# 5 Paints and finishes ............ 91

| | | |
|---|---|---|
| **5.1** | **Terminology, composition and applications** | **91** |
| **5.2** | **Substrate preparation** | **91** |
| **5.3** | **Priming** | **92** |
| **5.4** | **Mixing and application** | **92** |
| **5.5** | **Renovation coats** | **93** |

# 6 Dry earth construction ............ 95

| | | |
|---|---|---|
| **6.1** | **Introduction** | **95** |
| **6.2** | **Clay panels** | **95** |
| **6.3** | **Dry stacked walling** | **98** |
| 6.3.1 | Dry stacked wall infill | 99 |
| 6.3.2 | Dry stacked wall lining | 99 |
| **6.4** | **Ceiling overlays and ceiling and roof infill** | **100** |
| **6.5** | **Material and building element properties** | **103** |

| 6.5.1 | Mechanical properties | 103 |
|---|---|---|
| 6.5.2 | Thermal insulation, thermal capacity and vapour diffusion resistance | 104 |
| 6.5.3 | Sound insulation | 104 |
| 6.5.4 | Fire performance | 106 |

## 7 Internal insulation with earth materials ... 109

| 7.1 | **Internal insulation in general** | 109 |
|---|---|---|
| 7.1.1 | Introduction and background | 109 |
| 7.1.2 | Requirements of building materials for internal insulation | 110 |
| 7.1.3 | Suitability of earth building materials for internal insulation | 111 |
| 7.1.4 | Dimensioning internal insulation | 111 |
| 7.1.5 | Preparing existing walls for internal insulation | 113 |
| 7.1.6 | Junctions with walls and ceilings, and window and door reveals | 114 |
| 7.1.7 | Bearings for timber joists | 115 |
| 7.1.8 | Minimising air leakages with internal plastering | 116 |
| 7.2 | **Light earth wall linings (wet construction)** | 117 |
| 7.2.1 | Description of the internal insulation | 117 |
| 7.2.2 | Constructing light earth walls | 117 |
| 7.2.3 | Types of light earth and specific aspects of their construction | 120 |
| 7.2.4 | Building duration and drying | 121 |
| 7.2.5 | Fixing items to the light earth wall lining | 122 |
| 7.3 | **Light earth masonry wall linings** | 123 |
| 7.3.1 | Description of the internal insulation | 123 |
| 7.3.2 | Constructing masonry wall linings and appropriate materials | 123 |
| 7.4 | **Insulation board wall linings** | 125 |
| 7.4.1 | Description of the internal insulation | 125 |
| 7.4.2 | Mortar layer, application and fixing of the insulation | 126 |
| 7.4.3 | Kinds of insulation boards | 127 |
| 7.5 | **Material and building element properties** | 129 |
| 7.5.1 | Thermal insulation and protection against moisture | 129 |
| 7.5.2 | Sound insulation | 130 |
| 7.5.3 | Fire performance | 131 |

## 8 Earth block masonry ... 133

| 8.1 | **Introduction** | 133 |
|---|---|---|
| 8.2 | **Earth blocks** | 134 |

| | | |
|---|---|---|
| 8.2.1 | Base material and manufacturing methods | 134 |
| 8.2.2 | Requirements of earth blocks | 134 |
| 8.2.2.1 | Usage Classes | 134 |
| 8.2.2.2 | Inner and outer geometry | 135 |
| 8.2.2.3 | Bulk density and bulk density classes | 138 |
| 8.2.2.4 | Compressive strength and deformation of earth blocks under load | 138 |
| 8.2.2.5 | Moisture and frost resistance | 139 |
| 8.2.2.6 | Fire performance | 142 |
| 8.3 | **Earth masonry mortar** | 143 |
| 8.4 | **Non-loadbearing earth block masonry with & without timber studs** | 144 |
| 8.5 | **Loadbearing earth block masonry** | 146 |
| 8.5.1 | General aspects | 146 |
| 8.5.2 | Construction principles | 147 |
| 8.5.3 | Loadbearing structure and dimensioning | 147 |
| 8.5.4 | Physical performance of loadbearing earth block walls | 149 |
| 8.6 | **Material and building element properties** | 149 |

## 9 Rammed earth construction ............... 151

| | | |
|---|---|---|
| 9.1 | **Introduction** | 151 |
| 9.2 | **Rammed earth** | 152 |
| 9.2.1 | Raw materials and manufacture | 152 |
| 9.2.2 | Properties | 153 |
| 9.2.2.1 | Bulk density | 153 |
| 9.2.2.2 | Measure of shrinkage | 153 |
| 9.2.2.3 | Compressive strength and modulus of elasticity | 153 |
| 9.2.2.4 | Moisture and frost | 154 |
| 9.2.2.5 | Fire performance | 154 |
| 9.3 | **Constructing rammed earth walls** | 155 |
| 9.3.1 | Introduction | 155 |
| 9.3.2 | Building material preparation | 155 |
| 9.3.3 | Shuttering and formwork | 156 |
| 9.3.4 | Material insertion and compaction | 158 |
| 9.3.5 | Removal of formwork and touching up | 160 |
| 9.3.6 | Drying | 160 |
| 9.4 | **The design of rammed earth walls** | 161 |
| 9.4.1 | Constructional measures for weather protection | 161 |

| | | |
|---|---|---|
| 9.4.2 | Embedded elements | 163 |
| 9.4.3 | Reinforcement | 164 |
| 9.4.4 | Installations | 166 |
| 9.4.5 | Internal finishes of rammed earth wall surfaces | 166 |
| 9.5 | **Non-loadbearing rammed earth walls** | 167 |
| 9.6 | **Loadbearing rammed earth walls** | 167 |
| 9.7 | **Rammed earth floors** | 168 |
| 9.8 | **Physical properties of rammed earth** | 170 |
| 9.9 | **Building element properties** | 171 |
| **10** | **Renovation – historical earth buildings** | **173** |
| 10.1 | **Introduction** | 173 |
| 10.2 | **Solid earth wall constructions** | 174 |
| 10.2.1 | Weller earth construction | 174 |
| 10.2.1.1 | Weller earth and its properties | 177 |
| 10.2.1.2 | The construction of weller earth buildings | 179 |
| 10.2.2 | Historical rammed earth construction | 187 |
| 10.2.2.1 | Rammed earth and its properties | 190 |
| 10.2.2.2 | The construction of rammed earth buildings | 191 |
| 10.2.3 | Earth brick construction | 194 |
| 10.2.3.1 | Earth bricks and earth masonry mortar | 196 |
| 10.2.3.2 | The construction of earth brick buildings | 197 |
| 10.2.4 | Damages and repair of massive earth constructions | 199 |
| 10.2.4.1 | Material damage and reduced cross-section as a result of rising damp | 200 |
| 10.2.4.2 | Damages to plaster and render, weathering and washing out | 203 |
| 10.2.4.3 | Cracks | 206 |
| 10.2.4.4 | Pest infestation | 208 |
| 10.2.5 | Thermal insulation | 208 |
| 10.2.6 | Building material and building element properties | 209 |
| 10.2.6.1 | Mechanical properties | 209 |
| 10.2.6.2 | Selected physical properties | 210 |
| 10.3 | **Timber-frame panel infill** | 211 |
| 10.3.1 | Panel infill techniques | 214 |
| 10.3.1.1 | Wattle with straw-clay daub | 214 |
| 10.3.1.2 | Staves with straw-clay | 218 |
| 10.3.1.3 | Panel infill with earth brick masonry | 220 |

| | | |
|---|---|---|
| 10.3.1.4 | Internal and external facing coats | 220 |
| 10.3.2 | The repair of external timber-frame panels | 223 |
| 10.3.2.1 | The repair of panels made of wattle & daub or staves & straw-clay | 223 |
| 10.3.2.2 | The repair of panels made of earth brick masonry | 225 |
| 10.3.3 | New panel infill | 226 |
| 10.3.3.1 | New panel infill with wattle and daub or staves and straw-clay | 226 |
| 10.3.3.2 | New panel infill with earth brick masonry | 227 |
| 10.3.4 | External render | 231 |
| 10.3.4.1 | Exposed timber-frame constructions and weathering | 231 |
| 10.3.4.2 | Rendering timber-frame panels | 233 |
| 10.3.4.3 | Rendering and cladding entire façades | 237 |
| 10.3.5 | Building material and building element properties | 238 |
| 10.3.5.1 | Mechanical properties | 238 |
| 10.3.5.2 | Thermal insulation and water vapour diffusion resistance | 239 |
| 10.3.5.3 | Sound insulation | 240 |
| 10.3.5.4 | Fire performance | 240 |
| 10.4 | **Timber beam ceiling infill** | 242 |
| 10.4.1 | Traditional ceiling infill techniques | 242 |
| 10.4.1.1 | Staves with straw-clay infill | 242 |
| 10.4.1.2 | Earth reels | 244 |
| 10.4.1.3 | Ceiling insert with loose or compacted earth infill | 246 |
| 10.4.2 | Repair of old ceiling infill | 248 |
| 10.4.2.1 | Repair of ceiling infill made of staves and straw-clay or earth reels | 248 |
| 10.4.2.2 | Repair of timber boarding with loose or compacted fill material | 248 |
| 10.4.3 | New ceiling infill | 248 |
| 10.4.3.1 | New ceiling infill made of staves and straw-clay or earth reels | 248 |
| 10.4.3.2 | New ceiling inserts with earth mass or earth loose fill material | 249 |
| 10.4.4 | New ceiling plaster | 250 |
| 10.4.4.1 | New ceiling plaster on ceilings with staves & straw-clay or earth reels | 250 |
| 10.4.4.2 | New ceiling plaster on ceiling inserts with earth fill material | 252 |
| 10.4.5 | Building material and building element properties | 252 |
| 10.4.5.1 | Mechanical properties | 252 |
| 10.4.5.2 | Thermal insulation | 252 |
| 10.4.5.3 | Sound insulation | 253 |
| 10.4.5.4 | Fire performance | 254 |
| 10.5 | **Earth floors** | 255 |
| 10.5.1 | Historical earth floors | 255 |

| 10.5.2 | The repair of historical earth floors | 256 |
|---|---|---|
| 10.6 | **Historical earth plasters** | 257 |
| 10.6.1 | Description of historical earth plaster methods | 257 |
| 10.6.2 | The repair of historical earth plasters | 259 |
| 10.6.3 | Building material and building element properties | 260 |

## 11 Building legislation and business practice — 261

| 11.1 | **Earth building regulations** | 261 |
|---|---|---|
| 11.2 | **Building trades and earth building qualifications** | 263 |
| 11.3 | **Calculating the cost of earth building works** | 264 |
| 11.3.1 | Typical work times | 265 |
| 11.3.2 | Calculating typical constructions | 272 |

## Bibliography — 275

## Index — 283

# Foreword

Earth is a natural material that in recent years has advanced to become a high-quality building material for modern building projects. Its aesthetic qualities and character along with its beneficial effect on the indoor climate and general well-being are widely recognised. Of particular relevance are its environmental properties, for example the unparalleled low energy balance of many earth building materials. As an authentic historical building material, the physical characteristics of earth are also highly prized for the conservation and renovation of historic buildings.

This book is the product of many years' practical experience of earth building in Germany, where a traditionally hand-made material has had to prove itself in the context of a highly industrialised building sector. As a consequence, earth building applications and products in Germany are of a particularly high standard. "Earth Building Practice" offers architects, engineers and building tradesmen from around the world the opportunity to benefit from the experience of earth construction in Germany. The information presented in this book is expressly directly towards the needs of these users. Rather than presenting current research findings, it is our intention to provide concrete information and guidance for planning and construction practice that already incorporates the current state of research.

The chapter on *the raw material – soils for construction* provides a general and comprehensible understanding of the raw material with additional detail on aspects of particular relevance. It introduces a new proposal for a simple means of testing shrinkage and compression with a view to assessing different soil compositions and their potential properties as a building material. The technical characteristics of relevant *earth building materials* are presented alongside those of other typical building materials.

In this book particular attention has been given to two main areas of focus. The first of these is *earth and clay plasters*, the field in which earth building materials are currently most widely used. We describe in detail the processing of plasters with machines as well as working with coloured plasters and coatings.

The chapter on *dry earth construction* details a comparatively modern construction technique that has seen considerable developments in recent years. A particularly important application is the use of earth materials as *internal insulation* due to the need to improve the energy performance of existing buildings. Earth building techniques are well suited for such applications due to their ability to adapt to the conditions of the existing building fabric.

The chapter on *earth block masonry* reflects the most recent state of the art as detailed in the current norms under development. Similarly, modern *rammed earth construction* is examined in the context of its increasing application over the last 15 years and its advancement to a popular means of architectural expression.

The second main area of focus is the *renovation of historical earth building structures*. Christof Ziegert obtained his doctorate on the historical tradition of "weller" construction, Germany's equivalent to cob, and for many years has studied monolithic earth constructions around the world, and in the former East Germany in particular. Ulrich Röhlen has worked for two and a half decades in the field of renovating earth and timber-frame constructions. The chapter contains a comprehensive description of historical earth building techniques along with detailed best practice guidelines for the repair of earth building constructions.

The final chapter on *building legislation and building practice* describes the legislative framework governing standards as well as trade skills in Germany. The German system is regarded around the world as being stringent and oriented towards maintaining high quality standards. As such, it serves as a basis for similar developments in other countries. Finally, Earth Building Practice closes with an up-to-date overview of typical work durations for different earth construction techniques and a side-by-side comparison of building costs for typical construction elements made with earth as well as other building techniques common in Germany.

Ulrich Röhlen, Christof Ziegert
Berlin, Düsseldorf, September 2010

# 1 Earth building today

## 1.1 The state of the art of earth building materials

Earth is a building material that has been used throughout the world for thousands of years. It is also a modern building material. Earth is well suited for use in many of today's most important applications in the building sector such as the renovation of historic buildings, new energy-efficient construction or for building in developing countries.

That building with earth is once again widely practiced in Germany is the product of ongoing quality-oriented initiatives in the fields of training, planning, publicity and the development of norms over the past decades, and is not least thanks to the properties of the building material itself.

The incorporation of the *Lehmbau Regeln*, the German earth building codes, into the respective building regulations of the majority of the German federal states represents a milestone in securing the status of earth building in modern building practice. The *Lehmbau Regeln* [DVL, 1998] were drawn up by the German Association for Earth Building, the Dachverband Lehm e.V. (DVL) and were subject to technical approval by the German Institute of Building Technology (Deutsche Institut für Bautechnik, DIBt). Since 1998 it has been possible to use earth building materials for certain, clearly-defined loadbearing and non-loadbearing applications without the need to obtain special consent in each individual case from the building control authorities. The third revised edition of the *Lehmbau Regeln* [DVL, 2009] was approved by the DIBt in early 2008. This edition reflects the knowledge gained over ten years of earth building practice as well as in the application of the codes themselves. At the time of writing, draft norms are being drawn up for industrially manufactured earth building materials (earth blocks, earth plasters and earth mortars). This addresses the DIBt recommendation that these product groups should in future be regulated by DIN norms. The next stage will be the development of European norms.

The degree of industrial prefabrication and quality standards of earth building materials is on a par with other conventional building products. Typical products include dry ready-mix mortars delivered in sacks or silos, earth building boards and panels or large-format earth blocks. Material-specific product forms such as naturally-moist, ready-mixed building materials made of non-dried excavated earth and delivered on site in "big-bags" unite environmentally-friendly manufacturing methods with modern-day delivery requirements. The *Lehmbau Regeln* do, however, permit the traditional manual mixing of earth building materials, ideally using excavated soils sourced directly from the building site and tested for their suitability.

While earth building has become more widespread in Germany as well as in other industrialised nations over the past two decades, the thickness of the layers that are applied has grown successively thinner. The market segment of clay and earth plasters, and coloured fine-finish plasters in particular, has grown disproportionately. This development is, it seems, a response to the growing social acceptability of earth as a building material. The improved perception of earth has in turn benefitted other earth construction methods, even monolithic load-bearing construction.

In Germany, unlike in many other regions of the world, almost all earth building materials and techniques *do not employ stabilising additives*. Through the careful choice of suitable soils and appropriate additives it is possible to manufacture high-quality building materials that do not contain additional binding agents and fulfil the demands of most typical and appropriate applications of earth building materials, e.g. surfaces protected from the weather with low strength requirements. In Germany, it is not so much that there is little will or necessity to use earth building materials for inappropriate applications but rather a desire to employ a natural, pure and unadulterated building material.

## 1.2 Reasons for using earth building materials

The renaissance of earth building in the 1980s can be attributed almost exclusively to the excellent environmentally-friendly credentials of the building material. Today, if one were to ask the now much broader spectrum of clients and architects who specify earth materials for the reasons behind their choice, many will cite aesthetic reasons or the improved indoor air climate. It is largely because earth embodies all these qualities that the renewed popularity of earth building, sparked off perhaps by the Zeitgeist of the early 1980s, is still going strong some 30 years later.

### Environmentally friendly

The extraction, manufacture and processing of many building materials often involves serious interventions in the natural environment, the production of emissions and the consumption of large amounts of resources. Every responsible planner should aim to reduce the damaging effects of such processes on our natural living environment. Traditional natural materials such as wood, earth and stone need only be extracted from the natural life cycle and can be directly worked or processed. Unlike other binding agents, the cohesive binding effect of clay in earth building materials does not need to be activated by firing or chemical curing. And because the binding forces in clay are reversible, earth building materials can be replasticised for reshaping or reusing in another form. Other building materials such as steel and glass can also be reshaped but require the renewed input of large amounts of energy.

These key differences to other kinds of building materials become particularly apparent in a comparative assessment of sustainable building material and construction criteria. For example, the equivalent $CO_2$ of naturally-moist earth plaster delivered on site is just 5% of that of a cement plaster. At present, extensive test series are being undertaken to determine reliable values for the key characteristics of different earth building materials with a view to making these available through relevant databases.

### User comfort and health

According to the Federal German Health Agency, almost 2 million people in Germany suffer from some form of physical or mental disorder caused by indoor environments. When converting or constructing buildings, it should be a primary aim to create pleasant, comfortable and healthy living and working environments. An assessment purely on the basis of standard criteria such as structural stability, fire safety and noise and thermal performance does not go far enough. Hygiene and humantoxological aspects are becoming ever more important, a product of the reduced air change rates in well-insulated and sealed buildings.

With lower air change rates, the role played by the enclosing surfaces of the room becomes more important. This applies particularly with regard to indoor air humidity. To avoid widely fluctuating moisture levels in critical areas, it is important that the outermost surfaces of the wall are able to accommodate fluctuations in humidity, caused for example by showering, cooking and heating, by absorbing a degree of the airborne moisture. Excess moisture can later be released back into the room when it is ventilated. Conversely, where the indoor air is too dry, moisture can be retained within the room. This material property, known as sorption capacity, does not replace the need for ventilation but improves the hygrothermal conditions in indoor spaces, particularly where there is a low air change rate. The water vapour sorption capacity of earth building materials exceeds that of other building materials by a large margin. They are therefore ideally suited for improving the indoor room climate.

The extremely large internal surface area of the clay minerals, and therefore also of the earth building materials, serves not only as an effective buffer for airborne moisture, but also contributes to a limited but nevertheless measurable degree to the sorption of smells and odours.

Tests have shown that earth building materials of sufficient thickness (≥24 cm) are able to dampen pulsed high-frequency waves (for example as used for mobile telephone telecommunications) much better than other building materials.

### Aesthetic aspects

Outstanding examples of earth architecture such as the rammed earth walls of the *Chapel of Reconciliation* in Berlin or the earth plaster surfaces of the *Kolumba Museum* in Cologne, recipient of the German Architecture Award in 2009, contribute towards a general change in the perception of earth building. Experience of such buildings quickly dissipates preconceived notions of a "brown and crumbly" material and demonstrates that earth surfaces can be contemporary, high-quality and even exquisite; something that the Japanese have known for thousands of years.

On the other hand, the legacy of traditional earth building constructions all over the world, some of which bear testimony to the above preconceptions, exhibit a powerful presence, as reflected by the large number of earth constructions on the World Heritage List.

### Constructional aspects

Earth is most definitely not a "faster-higher-further" building material. Thankfully, these are not the only kinds of construction in the building industry, and if anything declining. Instead, properties that influence aspects such as building physics and building preservation are becoming increasingly important so that buildings become more sustainable and robust in the event of faults. By way of example, earth building materials offer a solution to the complex problem of insulating the internal face of historical buildings through their capacity to both transport as well as retain moisture on a long-term basis.

### Global aspects

Compared with the building tasks facing developing countries and emerging economies, our problems seem relatively small. The lasting improvement of living conditions in such countries can only be achieved by appropriately improving and developing traditional building methods – which in many cases employ earth. The modern renaissance of earth building in Germany and Europe is both encouraging and important precisely because developing regions are adopting the building materials and methods of industrialised nations. In ideal circumstances this could have the potential to reinforce regional building traditions and with it local cultural identity.

# 2 The raw material
## – soils for construction

## 2.1 Introduction

To manufacture, prepare and correctly apply a building material for a specific purpose requires a fundamental understanding of the raw material out of which it is made. In the case of earth building materials, this means above all an understanding of the binding agent in earth: the clay minerals.

As with all malleable masses used in construction, earth building materials consist of a binding agent and aggregates. The properties of the binding agent and of the aggregates as well as their relative proportions determine the properties of the malleable mass.

Clay, as the natural binding agent in earth, occurs in soils in many different forms with widely varying properties. They are almost always present in a mixture with non-binding sandy or stony constituents. Soils where the clay fraction is less than that of the granular constituents are known as *earth*.

Only certain earth mixtures are suitable for construction purposes. Mixtures with particularly pronounced binding characteristics (a factor of the kind and quantities of clay minerals present) are known as *rich* mixtures, those that are not very cohesive as *lean* mixtures. It is rare to find naturally-occurring earth mixtures with ideal relative proportions of binding and non-binding constituents.

In modern earth construction the use of factory-premixed products is widespread. Nevertheless, builders as well as architects may often be asked by the client whether the soil on site is suitable for making the required earth building materials. With a little experience it is possible to determine the suitability of a soil using simple field and laboratory tests and to identify a range of mixing recipes for manufacturing the resulting earth building material. More complex laboratory tests are usually only necessary for fine-tuning the material properties.

## 2.2 Types of soils and their formation

The types of rock formed in the earth's crust such as granite, gneiss or basalt are subjected at the earth's surface to chemical and physical weathering processes. These processes gradually remould the fabric of the rock fabric and alter their mineral structure. Unstable mineral phases such as feldspar and mica disintegrate into their more stable mineral components. A by-product, or end-product, of this process is the formation of cohesive and non-cohesive minerals: clay minerals as well as oxides and hydroxides of iron, aluminium or magnesium and quartz. In the terminology of soil mechanics, one speaks of earth when these minerals are present in the ground in particular quantities, kinds and grain sizes. The weathering processes are as old as the rock itself. The alteration and formation of clay minerals and earths is an ongoing process that takes place constantly, not only in rocks but also in the earth and even in the earth surfaces of buildings that are exposed to the elements.

Different natural processes that take place at the earth's surface can transport soils from their primary place of deposition, for example at the foot of a rock massif, to other locations. *Mountain soils* and *slope wash* that are transported by water are known as *fluvial (river)* or *alluvial (floodplain) soils*, those that are transported by glacial action as *glacial* or *moraine soils* and those that are transported by wind as *loess soils*. The different transportation mechanisms also result in selection processes in the soil. *Mountain soils* and *slope wash* contain a broad spectrum of freshly crushed grains. The proportion of very fine binding and non-binding grains is relatively small. The grain size distribution of *fluvial* or *alluvial soils* are predominantly determined by the changing flow velocities of the water. Depending on the place of deposition, earth deposits are typically to be found in several strata with different compositions, often veined with bands of sand and gravel. *Glacial soils* that date back to the Ice Age are often more or less uniformly mixed with rounded grains of different granularity. The grain size distribution of wind-borne *loess soils* have the least spread of grain sizes. Such soils often have a very high proportion of *silts* or so-called *fines*. Due to their natural homogeneity, they are often easy to work.

## 2.3 Clay minerals – the binding agent

### 2.3.1 Structure and cohesion

Clay minerals occur naturally in a wide variety of different forms. The following description is, therefore, necessarily somewhat simplified. Clay minerals are silicates with a lamellar structure with a size ranging from ca. 0.1 to 4 micrometre (μm). The size and respective chemical composition are a product of the kind of rock and degree of weathering.

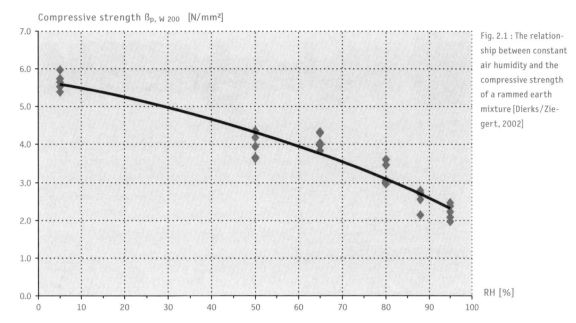

Fig. 2.1 : The relationship between constant air humidity and the compressive strength of a rammed earth mixture [Dierks/Ziegert, 2002]

Depending on the kind of clay minerals, two or three tetrahedral or octahedral-shaped layers of minerals join to form sheets. These so-called two or three layer clay minerals have different structures and accordingly differ in terms of their properties, sometimes considerably (see sections 2.3.2 and 2.3.3).

The binding force of the clay minerals can be attributed to localised differences in charges that occur at the surfaces of the sheets. In contrast to other binding agents used in construction, where hardening is a product of chemical curing, the *cohesive effect* of clay minerals derives primarily from the physical attraction of the particles.

Clay minerals bind water in crystalline form within the lamellae. This water content is not dependent on the ambient climatic conditions and is generally only released at high temperatures. The expulsion of crystalline bound water is non-reversible and modifies the original mineral structure. The exposing of clay-bonded masses to high temperatures is called *firing*.

As with all porous materials, condensation processes also occur in the pore structure of clay-bonded building materials in response to ambient humidity levels. This uptake of moisture is reversible and is known as water vapour sorption. Alongside the sorption of moisture through the pores, local differences in charges between the inner and outer surfaces of the clay minerals attracts water molecules in the vicinity. Accordingly a proportion of the water vapour sorption capacity of clay minerals can be attributed to its mineralogical composition. In contrast to the crystalline-bound water, water bound (weakly and reversibly) in the clay through sorption does affect the binding characteristics of the clay minerals. The binding force decreases as more and more water molecules are bound between and within the structure of the clay

Fig. 2.2 : A building defect has caused the earth to replasticise and subsequently harden on the external wall of a historical earth barn (view looking up at the eaves)

minerals. When the water molecules are released, the binding force becomes stronger again. The strength of an oven-dried earth building material is therefore the maximum strength that an unfired clay-bonded material can achieve. For example, tests conducted with rammed earth have shown that the strength of a sample in normal indoor air conditions (23 °C, 50 % RH, norm climate according to DIN 13279-2) is already 20 % lower than its potential maximum strength when oven-dried [Dierks/Ziegert, 2002]. As the oven-dried state is not relevant for real-life construction purposes, the reference strength of earth building materials is given for precisely this norm climate. If the relative humidity increases to 95 %, the material strength is only 50% of its potential maximum in its oven-dried state (figure 2.1).

Nevertheless, even at high levels of relative humidity, clay or earth materials are not in danger of softening to a malleable state in which they can be reshaped. This state requires the addition of water; water vapour is not sufficient. Large quantities of water cause the minerals to disperse and swell to a lesser or greater degree. The mass becomes malleable but still cohesive and hardens again after drying (figure 2.2).

The effect of water on the binding properties of clay has direct consequences for how we use clay-based building materials and the buildings we construct with them, for example their behaviour when exposed to the weather, damages that may result from rising damp in the assessment of building conditions. By way of example, the speed at which a two-storey rammed earth construction can be built is conditioned by the drying process across the thickness of the cross-section, as the tension in the cross-section increases the higher the wall becomes. Figure 2.3 shows how the compressive strength of a rammed earth sample increases as it dries from its moisture level during

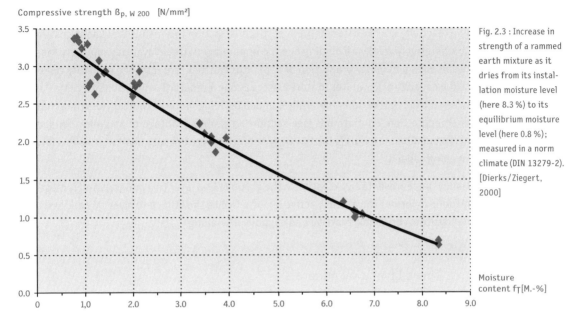

Fig. 2.3 : Increase in strength of a rammed earth mixture as it dries from its installation moisture level (here 8.3 %) to its equilibrium moisture level (here 0.8 %); measured in a norm climate (DIN 13279-2). [Dierks/Ziegert, 2000]

construction (here 8.3 %) to its equilibrium moisture level under normal climatic conditions (here 0.8 %).

A particular characteristic of clay materials, compared with other minerals, is their very large specific surface area as well as their so-called *intercrystalline reactivity*, which is most pronounced in three layer clay minerals. This describes their ability to temporarily or permanently bind or attach other substances in the mineral structure (*cation exchange capacity*) in response to changing environmental conditions. This applies both to water molecules as well as organic and inorganic compounds [Jasmund/Lagaly, 1993]. Alongside the pronounced capacity of clay minerals to buffer humidity, the intercrystalline reactivity also explains its limited but nevertheless noticeable and measurable capacity to bind odours [Müller, 2007].

The colour of clays is mostly determined by the presence of metal oxides and hydroxides, e.g. from iron, manganese and aluminium, that are present as impurities in almost every natural clay. The colour of a clay is not an indication of its properties.

### 2.3.2 Two layer clay minerals

The group of *two layer clay minerals*, which are also known as the *kaolinite group*, consist of minerals with 2 linked layers – one octahedral and one tetrahedral sheet. A kaolinite clay mineral consists of between 20 and 100 such minerals in stacked lamellae. Kaolinite particles are relatively large with a mean particle size of between 0.5 and 4 µm. The specific surface area of approx. 10 m²/g is relatively small compared with three layer clay minerals. The interlayer charge difference as well as the intercrystalline reactivity are likewise small. For earth construction this means that kaolinites are comparatively weak as a binding agent compared with three layer clay minerals. Materials with a majority proportion of kaolinites can lack strength. The humidity regulating capacity of kaolinites to store water vapour temporarily in the building material is larger than that of other minerally-bonded building materials but still less than that of smectite-bonded materials (see section 2.3.3). An advantage of kaolinite-bonded earth building materials is the low degree of shrinkage because kaolinites have a low swelling capacity.

Halloysite represents an exception in the group of two layer clay minerals and is mostly found in former volcanically active regions. Unlike the other two layer clay minerals, halloysite is very small, reactive and capable of swelling.

### 2.3.3 Three layer clay minerals

The elemental mineral group of *three layer clay minerals* consists of three layers of minerals with an octahedral sheet sandwiched between two tetrahedral sheets.

Within the group of three layer clay minerals, there are both minerals with and without swelling characteristics. In the former case, the addition or removal water causes the material to swell and contract. The most important groups of expandable three layer minerals are *montmorrilonite* and *beidellite*, while the most important group of non-expanding three layer clay minerals are *micaceous* minerals, of which the most prominent is *illite*.

The mean particle diameter of three layer clay minerals is usually smaller than 0.2 μm and therefore much smaller than two layer clay minerals. Figure 2.4 shows clearly the difference in particle size seen through a *scanning electron microscope* (SEM), with *kaolinite* on the left and *smectite* on the right.

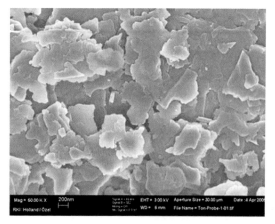
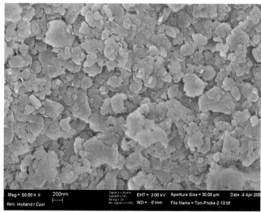

Fig. 2.4 : Two layer clay minerals (left) and three layer clay minerals (right) seen under a scanning electron microscope (SEM)

The specific surface area of three layer minerals can be as large as 1000 m²/g. The much larger surface area compared with two layer clay minerals is a product of the accessibility of the interlamellar spaces for reaction processes [Jasmund/Lagaly, 1993]. The lamellar surfaces of the three layer clay minerals exhibit large charge differences which means that the binding force of the three layer clay minerals is much larger than that of their two layer cousins. The intercrystalline reactivity is likewise many times higher than that of two layer clay minerals. As a result more water can be attached to the mineral structure, which leads to greater water vapour sorption characteristics. The swelling and contraction behaviour of earth materials made with three layer clay minerals has to be carefully compensated for by the addition of mineral and organic additives.

## Clay mineral analysis

Given the varying kinds and characteristics of clay minerals, it is clear that earth materials cannot be assessed solely on the basis of the absolute clay content.

The ability to identify the kind of clay mineral is likewise difficult because of their microscopic size. This requires the experience of a specialist mineralogist.

Clay minerals can be determined most accurately using *x-ray diffraction*. With this method it is possible to determine the kind of clay mineral as well as an approximate indication of their relative distribution. For the analysis, special complex-to-produce samples need to be prepared known as texture samples. *Differential thermo analysis* (DTA) and *thermogravimetry* (TG) as well as *infra-red spectroscopy* (FTIR) can also be used to ascertain further information about the clay minerals identified using x-ray diffraction.

To clarify morphological differences, *scanning electron microscopy* (SEM) or *transmission electron microscopy* (TEM) is used. Both methods can also be used for microchemical analyses, providing information about the chemism of the clay minerals. The preparation of samples both for SEM as well as TEM is complex and the analysis itself is time-consuming.

The method used in the ceramics industry to determine the expandability of clay minerals, the so-called methylene blue method, is of limited use for earth construction as the results of the test are only reliable for samples with an unusually high clay content. Meythylene blue is a special stain that is absorbed by the cation exchange capacity of the clay minerals. By examining the quantity of methylene blue that is absorbed, one can estimate the swelling capacity of the minerals.

## 2.4 Testing the suitability of earth mixtures

In addition to the kind and quantity of clay minerals it contains, the suitability of an earth mixture for making earth building materials depends on numerous other factors. As a result, the builder or architect needs access to sufficiently good means of evaluating the suitability of earth mixtures. One can differentiate between simple *field tests* which can be carried out in situ and *laboratory tests*.

An assessment of suitability should clarify the following questions:
- Is the earth mixture sufficiently cohesive to acquire the desired strength and bind together the aggregates?
- In what quantities, grades, forms and surface characteristics are sandy or stony constituents present in the earth mixture? Will these affect the workability of the desired earth building material?
- Can the earth mixture be prepared with reasonable effort?
- What degree of drying shrinkage is to be expected?
- Does it contain any undesirable constituents?

While it is possible to modify unsuitable earth mixtures or to eliminate undesirable constituents through special processing, such processes are laborious and wherever there is a choice builders will opt for earth materials that can be modified with as little effort as possible. In previous centuries, it was common practice to employ only building methods that were possible with the locally available material.

### 2.4.1 In-situ soil testing

The assessment of earths for building purposes using simple *field tests* requires a degree of experience as well as sensitivity in the physical sense of the word. With a combination of the two it is possible to make a reliable assessment of the general suitability of a soil or earth mixture, to establish a sufficiently good classification (see table 2.1) and correspondingly to develop a range of appropriate mixing recipes. These findings can be verified by additional testing of the material properties. A soil sample should be subjected to a series of several different field tests as it is not always possible to clearly determine certain kinds of soils with certain tests. Most of the tests detailed below are also described in the DIN 4022-1:1987-9 "Classification and description of soil and rock".

An initial *visual assessment of the material deposits* should be used to establish the nature and regularity of the deposits. If the soil is present in thin bands or strata with different qualities, samples from the individual bands must be taken and blended for testing. Similarly, it is important to determine whether the respective quality is present in sufficient quantities. A first indication of the shrinkage characteristics can be ascertained by looking for clod formation in dried sections of the material. Large clumps in a pit may mean that the material will be laborious to process.

The different field tests are conducted with different consistencies. In a comparison of the samples with one another, they should exhibit a similar consistency in the respective tests as this has a significant effect on the formability (plasticity) and the binding force (tensile strength in a viscoplastic state).

Taking a *naturally-moist sample*, one should try to form a ball out of the material (*ball test*). The ability to shape the material increases with the binding force of the earth. A very lean soil will disintegrate if dropped from a height of approximately 2 m onto a hard flat surface. A ball of lean to semi-rich earth may break apart but will not disintegrate. A ball of clay-rich earth will simply squash flat.

Cutting open a ball of naturally-moist earth with a knife (*cutting test*) and examining the resulting surface is a further simple test: a shiny surface indicates a rich mixture, a dull surface a lean mixture. Traces left by the knife indicate a sandy mixture.

The existence of humus (decomposed organic plant remains) in a naturally-moist sample can be determined by its characteristic smell (*smell test*). Earth that does not contain humus is more or less odourless. Small quantities of humus in an earth mixture are tolerable depending on the application. This is most critical for plastering where even small humus particles can cause surface discolourations ranging from small dark specks to patches of whitish salt efflorescence. The absence of humus does not mean that the sample is free of microbes: earth mixtures as well as sands, for example, have their own microbial flora [Röhlen, 2008]. As such they are no different to other raw materials used in construction. Only if the earth mixture is heat-treated as part of its later processing can the building material truly be termed sterile (excepting, of course, impurities that arise from its later use).

Both the binding force as well as the mineral granularity can be determined by rubbing a *viscous sample* between one's fingertips (*rubbing test*). Fine and sandy constituents can be felt between one's fingers. It is more difficult to detect the difference between silty and clayey constituents as non-cohesive or weakly-cohesive silty fines also feel soapy when rubbed between the fingers in a viscous state. When washing the sample off one's hands, however, silty soils wash off more quickly and easily (*wash test*). By taking a viscous sample and pressing together and then drawing apart one's fingertips, one can roughly estimate the cohesion of the sample by assessing its resistance to being pulled apart (stickiness) as well as the way it pulls apart. The more difficult it is to pull one's fingertips apart and the stringier the sample, the greater its cohesion.

Through a combination of crushing or breaking a *dry sample* or letting it fall on a hard surface one can acquire an impression of the strength that the earth mixture will have when used in construction. Samples with low binding strength can be crushed between one's fingers. Lean mixtures break into many small pieces, or even dust, when dropped. Likewise, if one rubs the surface of a dry sample, one can determine whether the material exhibits a basic level of resistance to abrasion.

### 2.4.2 Laboratory testing of earth mixtures

#### 2.4.2.1 Grain size distribution

A commonly used method to assess the suitability of earth mixtures for construction is to determine their *grain size distribution* according to DIN 18123, a technique commonly used in the field of soil mechanics. This is determined by assessing the mass distribution of different grain sizes by using a combination of *sieve and sedimentation analysis*. Figure 2.5 shows an example of the grain distribution curve of a loess soil.

According to the German geotechnical classification of soils, the boundary between the clay and silt fractions is given as 0.002 mm or 2 µm. All particles that are smaller than this value are termed the *clay fraction*; all particles that are larger are divided into the silt, sand, gravel and stone fractions. The term *clay fraction* is often taken to mean the proportion of clay minerals, i.e. of the cohesive constituents. From the mass fraction of the clay content, assumptions are then drawn regarding the high or low cohesiveness of an earth sample, or a high or low measure of shrinkage. But these conclusions are questionable for the following reasons:

▶ Kaolinite can exhibit particle sizes in excess of 2 µm but particles that are larger than 2 µm are no longer classified as being part of the clay fraction. That leads to an underestimation of the clay mineral content and of its binding force.

Fig. 2.5 : Grain size distribution curve of a loess soil determined with a combination of sieving and sedimentation

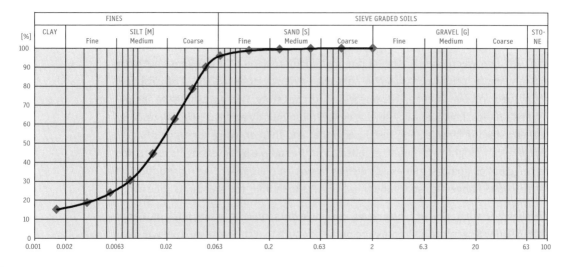

- Quartz, feldspars, carbonates as well as different metal oxides and hydroxides are also present in grain sizes of less than 2 μm. Because they have a much lower surface charge and a smaller specific surface area, they also have a much lower binding strength than clay minerals. According to *Heim*, the proportion of quartz in the clay fraction can be as much as 20–40 % [Heim, 1990]! That can in turn lead to an overestimation of the clay mineral content and the binding strength of the mixture.
- The binding strength of an earth mixture can vary considerably depending on whether the clay fraction contains two layer or expandable three layer clay minerals. It is, however, almost impossible to determine the mineral groups using sieve and sedimentation analysis. That leads in turn to false estimations of the binding strength, shrinkage measure, compressive strength etc.

As a means of establishing the suitability of an earth mixture for construction purposes, the grain size distribution is, therefore, not entirely suitable. It can nevertheless be useful for the manufacture of earth plasters and rammed earth mixtures in order to obtain an optimal distribution of grain sizes. With an overview of the grain distribution, it is possible to identify missing or under-represented fractions. As such the grain distribution can serve as a useful supplementary test procedure.

### 2.4.2.2 Binding force test (figure-8-shape test)

To assess the binding qualities of earth mixtures for earth building materials, the *binding force test* – with its distinct *figure-8-shaped* test specimen – is better suited than the quantitative analysis of the constituent grain sizes. This test assesses the sum of all the binding forces of all the clay mineral groups and the entire matrix. It is a little more complicated to undertake but provides quantitative results from which it is possible to characterise the earth mixture.

Using the binding strength test it is possible to measure the tensile strength of a naturally-moist to viscoplastic test specimen (figure 2.6; for the precise setup and procedure see [DVL, 2009]). The binding force or adhesion strength in its naturally-moist state is a key characteristic of earth building materials. Unlike cement or lime-bonded masses of similar consistency (i.e. before the chemical hardening process sets in), where the adhesion forces are weak and depend on the presence of water, the binding force of clay is already evident in its plastic state; only somewhat weaker than in its dry state.

After establishing the breaking stress from the breaking load applied in the test, the earth mixture can be clearly classified according to table 2.1. In addition the test values also provide a first indication of its suitability for the manufacture of certain earth building materials.

Fig. 2.6 :
Test rig for testing the binding force of earth mixtures (figure-8-shape test)

Earth mixtures with a breaking stress of less than 0.005 N/mm² cannot be differentiated sufficiently with this test and other tests are better for assessing their potential suitability. According to the *Lehmbau Regeln*, these are termed "generally unsuitable for construction purposes". There are some exceptions such as earths that already exhibit a suitable grain skeleton and can therefore be used as plaster mortar without the need for further additives, or earth mixtures with a high natural lime content (see section 2.4.2.4).

A disadvantage of the binding force test is the inability to precisely forecast the expected or achievable measures of shrinkage and compressive strengths. While observations show that lean earth mixtures shrink less than rich mixtures with the same moisture content, the spread of results is so broad that it is not possible to draw any precise conclusions. Similarly, the binding force test is not able to assess the contribution of the natural lime content to the strength of the sample. In some cases, earths that exhibit excellent quantities when processed to form earth building materials, are deemed unsuitable due to insufficient binding strength. Suitable testing procedures are instead those that assess the dry compressive strength, for example the dropped ball test with a dry, spherical test sample carried out in-situ or the compressive strength test undertaken in the laboratory.

Table 2.1 : Classification of earth mixtures according to binding force using the 8-shape test [DVL, 2008]

| attached mass at breakage for the entire cross-section [5 cm²] | attached mass at breakage per cm² | breaking stress | description |
| --- | --- | --- | --- |
| [g] | [g/cm²] | [N/mm², MPa] | |
| up to 250 | up to 50 | up to 0.005 | generally unsuitable |
| ≥ 250 to 400 | ≥ 50 to 80 | ≥ 0.005 to 0.008 | very lean |
| > 400 to 550 | > 80 to 110 | > 0.008 to 0.011 | lean |
| > 550 to 1000 | > 110 to 200 | > 0.011 to 0.020 | semi rich |
| > 1000 to 1400 | > 200 to 280 | > 0.020 to 0.028 | rich |
| > 1400 to 1800 | > 280 to 360 | > 0.028 to 0.036 | very rich |
| > 1800 | > 360 | > 0.036 | clay |

### 2.4.2.3 Determining the compressive strength and shrinkage of earth mixtures

As an alternative to the binding force test, the determination of the measure of shrinkage and compressive strength has proven useful for establishing mixing recipes and their achievable properties. The necessary mortar prisms, made according to DIN EN 1015-11 (l × b × h = 160 × 40 × 40 mm), are produced in a consistency as similar as possible to how they will ultimately be used: earth destined for use as plaster mortar is added to the prism mould in normal plaster mortar consistency; earth destined for use as a rammed earth screed is added to the mould in a naturally-moist consistency. This makes it possible to make relatively accurate assessments of the expected measure of shrinkage in the building, as well as of the expected compressive strength. The shrinkage test is especially simple and straightforward to prepare.

The measure of shrinkage of a sample in plastic consistency is some 20 - 50 % higher than that of a naturally-moist mixture due to the much higher water content. The compression is 5 - 25 % less due to the greater porosity.

From the measure of shrinkage of the test prisms and the desired measure of shrinkage for the respective building material (e.g. less than 2 % for earth plaster mortars) it is possible to derive the need for additives, and with some experience the quantity of additives required. The compressive strength measured with the test prisms of earth material provide an indication of whether the earth mixture has the potential to attain the required compressive strength for its later use. For example, the addition of a well-graded, angular grain skeleton can improve the likelihood of a stable earth building material, but because the absolute proportion of binding material sinks, it can also be necessary to add richer earth mixtures or powdered clay to particularly lean earth mixtures.

### 2.4.2.4 Qualitative determination of the natural lime content

The qualitative assessment of the *natural lime content* in earth mixtures for construction can be undertaken by applying drops of hydrochloric acid or alternatively concentrated vinegar. The greater the quantity of lime and the finer its presence in the test specimen, the stronger the resulting effervescence will be. The natural lime content in earth mixtures consists of more or less fine particles of limestone (calcium carbonate), depending on the kind of earth. These particles range from coarse grains in the case of fresh alluvial soils to millimetre sized particles in loess soils. Limestone exhibits a degree of solubility when exposed to water enriched with $CO_2$ (such as rain water). From this solution solid mineral structures can result as seen, for example, in the stalactites and stalagmites formed in caves. If an earth building material that contains very fine and evenly distributed limestone particles is stored for a long period of time under damp conditions, these particles can partially dissolve. When the material dries, limestone and calcite mineral structures form alongside the clay mineral bonds and

Figs. 2.7 a and b: Salt effluorescence visible on reprocessed earth from soaked demolition material (Jahili Fort, Al-Ain, UAE)

contribute to the strength of the material. The process described here should not be confused with the effects caused by the addition of reactive lime (calcium hydroxide) as a binding agent to an earth building material.

In contrast to the material's natural lime content, the addition of reactive lime can lead to significant interactions with the clay minerals and may possibly negatively affect the strength characteristics of the material.

### 2.4.2.5 Determining soil salinity

*Salts (chloride, sulphate and nitrate)* are present in every earth mixture. Above a certain concentration, salts can lead to building defects of damages. Most earth mixtures contain a non-critical level of salt concentration. Testing is nevertheless advised in certain circumstances:

- Earth excavated from pits near to coastlines,
- Earth from pits in the vicinity of intensive animal farming (manure pollution),
- Earth from pits that may be contaminated with road de-icing salt, and
- Earth material recycled from demolition material.

In an international context, this list should be extended to include steppe soils or earth from desert regions.

The quantitative determination of potentially damaging salts is usually undertaken using *ion chromatography* or *spectral photometry*. Traditional wet chemical methods such as *gravimetry* and *potentiometry* are also employed.

The level of salt content acceptable in earth mixtures for construction depends on the intended application (both the building material and building element) as well as the conditions of the respective construction. In the case of building elements subject to fluctuating moisture levels (for example through exposure to the elements or rising damp) the tolerable level of salts is much lower than for building elements that are constantly dry. This is because temporarily soluble salts are systematically transported from all layers to the evaporation zone which can result in a critical build-up of

Fig. 2.8 : Dark patches indicate hygroscopically induced moisture resulting from the use of earth blocks with high salt content. Right: measuring the moisture content (Jahili Fort, Al-Ain, UAE)

salt concentrations manifested as damp patches with salt efflorescence and surface disintegration. For building elements subject to fluctuating moisture levels, an overall anion concentration of up to 0.05 % by mass does not usually cause problems. The kind of salts present is also of importance. Readily soluble nitrates are generally more problematic than low-soluble sulphates. For building elements that are kept dry, an overall anion concentration of 0.10 % by mass should be taken as the upper limit.

One should also take into account that readily soluble salts increase the moisture content of the building material due to their hygroscopic properties. This can have considerable impact on material properties such as strength as well as colour (figure 2.8).

## 2.5 Processing

The term *processing* covers all the operations involved in preparing the raw material for use as a building material. The aim of processing is to ensure that the resulting material fulfils the requirements for its later use, e.g. that it is of a consistent quality and free of unwanted contamination and that the clay minerals within the earth building material are deflocculated so that they are able to fully function as a binding agent.

Soils are usually excavated from a pit and processed with the help of machinery. Mechanical processing can be augmented, or ideally even replaced entirely, with natural processes. This is, however, more time consuming.

The choice of appropriate processing methods depends on a large number of boundary conditions, not least the quality of the excavated soil itself.

### 2.5.1 Natural processing of earth mixtures

Natural processing methods are most commonly used when earth building materials are to be prepared from earth material available on site. A variety of different methods can be used:

#### Soaking

*Soaking* involves allowing the earth to rest in a wet condition. After a while the clay mineral particles begin to disperse in the mixing water, causing the bonds to weaken between the crystallites to such a degree that it can be easily worked into a homogenous mixture. The time it takes for the clay minerals to disperse depends on the size of any lumps of clay in the mixture as well as the kind and size of the clay minerals present in the material. Highly expandable clays exhibit a sealing effect that hinders the process of deflocculation, which can take several weeks, even months. In the case of lean and semi-rich earth mixtures, it is usually sufficient to soak them overnight.

#### Weathering (summer)

*Summer weathering* involves exposing the soil or earth mixture to fluctuating levels of ambient moisture – at the very least it should be naturally-moist. The bonds between the clay minerals weaken, allowing the earth mixture to be sieved or mixed more easily. Although not as fast or effective as soaking, the resulting earth mixture has a firmer, naturally-moist consistency, which is necessary for applications such as rammed earth.

#### Weathering (winter)

*Winter weathering* (wintering) involves exposing the soil or earth mixture to fluctuating levels of ambient moisture and the cycle of freezing and thawing. The mixture should not be allowed to be any drier than naturally-moist. The process is similar to weathering in summer, except that freezing is very effective at loosening the structure of the material, making it possible to deflocculate clay minerals in very inhomogeneous materials. Afterwards, the mass can be homogenised by simply mixing it. The size of the area for laying out the material for wintering depends on the expected frost penetration depth.

### 2.5.2 Mechanical processing

Mechanical processing encompasses a wide range of means of mechanical agitation as well as sieving.

## Crushing

The most widespread processing method used in the *ceramics industry* is *crushing*. In this process, naturally-moist earth material or clay is crushed between millstone-like rollers and pressed through a coarse sieve. This serves, on the one hand, to homogenise the composition and moisture content of the mass and on the other breaks down any stony constituents to a grain size that is adequate for most applications. Some brickworks (especially those that have specialised in the production of earth building materials) also offer crushed earth mass for sale alongside raw excavated material.

## Grinding

For the preparation of certain earth and clay products, the raw material is required in dry power form. The pre-dried raw material is ground in a pulverising mill which in terms of principle and machinery is similar to that of a flour mill. Depending on the grade of grinding, the resulting mass can contain stony particles of up to 4 mm in size, which can be too coarse for use in thin layers of earth plaster. The resulting powder is usually sold in sacks, larger textile big-bag or in silos.

## Mixing

The mixing process involves turning and mixing naturally-moist or wetter earth material in a mixer or agitator to homogenise the mixture. This means of processing is most effective for lean earth mixtures. In the case of clay-rich earth mixtures the clumps of clay may not be broken down sufficiently or only after lengthy mixing.

Where earth mixtures naturally exhibit a homogenous structure and are not very lumpy (due to their relatively low cohesion), which is often the case with loess soils, the processing and mixing in of additives can be undertaken in a single operation. Aside from sufficiently powerful milling machinery, this requires appropriate additives to break up the mixture (e.g. angular sand, crushed grains). The production of straw clay is about the limit of what is possible with paddle mixers.

## Whisking or slurrying

*Mixing* with a whisk-like blunger or *slurrying* involves mixing earth with water to a slurry-like mass. In many cases this process is used in advance of natural soaking. This process is very effective in dispersing the clay minerals. In addition the heavier, coarser stony constituents sink to the bottom of the mixing container, allowing one to separate them off. Due to the high water content of the mixture, this method is most suited to the production of earth plasters.

Sieving

If a mixture contains stony constituents that are too coarse, these must either be crushed or sieved. Naturally-moist earth can be *sieved* provided that the consistency is sufficiently dry and that clay-rich mixtures do not contain clumps. Where mixtures are stony or exhibit a tendency to clump, another processing method must first be used before the mixture can be sieved.

# 3 Earth building materials
## – Composition and properties

Raw excavated soil mixtures are only rarely suitable for use as a building material without the addition of additives. Some exceptions are, however, possible such as stony mountain soils which exhibit characteristics suitable for use as rammed earth.

Earth building materials consist in general of soil excavated from a clay or earth pit and aggregates. The kind and proportions of additives determine the mechanical and physical properties of the material as well as its appearance.

## 3.1 Aggregates and additives

Aggregates and additives are added to soils to modify their mechanical and physical properties for use as building materials. The boundary between the two kinds of additives is fluid and there is not a clear differentiation between the two.

### 3.1.1 Aggregates

Aggregates are materials that are bound or enveloped by the soil and can stem from mineral or organic sources. Together with the soil they constitute the mass of the unformed building material. Aggregates have little to no effect on the hardening characteristics of earth mixtures.

Mineral aggregates: Crushed stone – gravel – grit – sand

Stony and sandy aggregates are used to *lean down* a mixture. Leaning down involves the addition of coarser particles to a grain mixture consisting of predominantly fine particles. In soils for construction, the so-called 'fines' are made up of silts and clay. The inner surface area of the mixture per unit volume, that is the surface that will be enveloped with water when worked (plasticisation), sinks with the proportion of fine particles. The smaller the proportion of water, the fewer the voids that remain when the material dries. The fewer voids there are, the lower the stresses in the material that can lead to shrinkage cracking. Leaning down therefore minimises the degree of volumetric shrinkage during drying (*drying shrinkage*). In addition, the use of angular aggregates results in a more bulky matrix in which pores and small cavities form in the mixture without the overall volume shrinking and forming cracks.

The *workability* of strongly adhesive, viscous and sticky earth mixtures can be improved through the addition of sand.

The addition of stony and sandy aggregates can increase the strength of the mixture. As with mortars or concrete, earth building materials are strongest when the grain fractions are present in equal proportions. The proportion of the respective grain fractions can be plotted as a grain distribution curve (figure 3.1). The strength of a mixture can be improved through the addition of grain fractions that are lacking or insufficiently present in the available earth mixture. In addition, the form of the grains of aggregate influences the strength of the building material. Angular grains form a more stable structure than rounded grains. As the binding agent of the earth building material (clay) is comparatively weak, the nature of the grain constituents is particularly important.

The choice of the coarsest grain fraction is a factor of the intended dimensions of the resulting building element. For example, thick layers of plaster have a coarser grain distribution than thin skim coats to ensure the necessary stability.

An advantageous grain distribution helps to reduce the material's *sensitivity to water*, as the strength of the mixture is a factor of both the stable grain structure and the water-soluble clay mineral bonds. This applies likewise to the material's *sensitivity to frost*. Mixtures consisting solely of fine-grain building materials are more likely to crack apart due to frost than coarse grain mixtures.

Fig. 3.1: Grading curve for rammed earth

Stone and sand aggregates should not, however, be added in too large quantities. If the proportion of binding agent is too small, the strength of the material suffers.

The bulk density can be influenced by the kind of stone aggregate, for example in the case of rammed earth. With aggregates consisting of dense, low-porosity basalt, rammed earth can have a bulk density of 2400 kg/m³ instead of the typical 1800 - 2000 kg/m³.

Aggregates consisting of coloured grains additionally influence the appearance of the rammed earth and facing earth plaster surfaces.

### Lightweight mineral aggregates: expanded clay – pumice – perlite – foam glass

Fine grain perlite increases the capacity of mortars to retain water and accordingly influences their *workability*. This comparatively soft and spherical coarse aggregate also has a negative effect on the *strength* of an earth building material.

Expanded rock or other expanded mineral materials are used primarily to reduce the *bulk density* of the earth material, for example to lighten the weight of ceiling fill materials to a tolerable level for the structure. These aggregates also impede the thermal conductivity of the material; lightweight mineral earth mixtures are therefore suitable for use as internal insulation.

### Organic aggregates: straw – flax – hemp shavings– cellulose (fibres)

Like stony or sand aggregates, plant fibres are added to earth mixtures to *lean down* the mixture. The effect is the same as described above (section 3.1.1). Fibres are also able to sustain tensile forces, reducing the effect of *drying shrinkage*.

The addition of plant fibres also improves, or enables, the *workability* of the material. Accordingly, the panel infill of wattles or staves in timber-frame constructions is made possible through the use of straw clay mixtures and rolled earth reels are held together with long-stem straw. Similarly, fine plant fibres contribute to the pliability of earth plaster mortars, cellulose fibres even more so. Fine fibres also increase the water retention capacity of materials.

Plant fibres influence the strength of the material by forming a compound with the mineral mixture to create a stable overall structure. This in turn reduces the material's *sensitivity to water*. Earth blocks with fibre additives break apart much more slowly when exposed to water than blocks consisting solely of mineral material.

Fibres are able to take up tensile forces within the material and act as reinforcement. Long-stem straw helps make weller mixtures sufficiently cohesive for wall construction and the stability of clay panels and earth building boards is largely due to the addition of plant fibres. Fibres likewise help plaster mortars sustain the effect of tensile forces originating from the underlying substrate or from thermal expansion or contraction.

In addition to their reinforcing function, coarse fibres can also contribute to the stability of a material due to their compressibility. When heat causes the mineral constituents to expand, the soft fibres can act as a buffer. This property also reduces the material's *sensitivity to frost*.

The *bulk density* and in turn the thermal conductivity of a material mixture can be reduced through the addition of plant fibres. This is particularly desirable for the manufacture of lightweight earth blocks. For plaster mortars, the reduction in bulk density generally has little effect. Historical straw-clays, however, can be very lightweight and contribute in part to the thermal insulation.

The *sorption capacity* of earth building materials is influenced indirectly by the addition of plant fibres, as leaning down and reinforcement makes it possible to prepare earth mixtures with a high proportion of sorptive clay minerals. The sorption of the fibres themselves can also have a slight effect.

A high proportion of plant fibres can help soften the *acoustic properties* of a room as soft and open-pore surfaces reflect less noise than hard, closed surfaces.

In terms of *fire performance*, the addition of fibres can have a positive effect, as fibre-rich mixtures retain their stability better in the event of fire and exhibit a higher mechanical loadbearing capacity. The reduced thermal conductivity can also be advantageous in the event of fire. The flammability of the material is only affected when the proportion of fibres is very high (see below). In terms of fire performance classification, however, the addition of fibres has a negative effect as the *non-flammability* of the mineral material is compromised by the addition of combustible constituents.

Earth has its own *microbial flora*, as does sand. Plant additives can contain bacteria and mould that arise during the cultivation of grasses and cereals. At present, it is not clear whether and to what degree organic materials increase the risk of mould during the processing and drying of earth mixtures (see below). Microorganisms are generally to be found in traditionally manufactured earth building materials that are mixed and stored in a naturally-moist condition. As such, they are present in most kinds of historical buildings. If the individual constituents of the building material are dried before mixing, the concentration of bacteria and mould can be reduced significantly. Heat-treated (> 80 °C) products, like many other mineral-based building materials, can be regarded as practically free of germs apart from the odd impurity.

A material's *susceptibility to mould formation* during the drying period (wet earth constructions) may increase as a result of plant fibres, as these serve as suitable nutrients. Given the effect of other factors such as the duration of the drying period, ambient humidity levels and temperature, the relevance of this fact is unclear. Mould formation during the building phase can be avoided through appropriate preparation and drying precautions (section 4.5.6). Mould formation on earth surfaces containing plant fibres is not known to occur during the later use phase of the building.

The *appearance* of materials and surfaces are influenced significantly by the addition of plant fibres. A wide variety of different surface textures can be achieved through the addition of different aggregates.

### Wood aggregates: wood shavings – sawdust – wood chips

Wood shavings are added to earth to lean down and to improve the strength of the mixture. As with plant fibres, the material's *sensitivity to water and frost* is reduced. Wood chips are commonly used as an aggregate for lightweight earth mixtures. The effect of wood aggregates on the *bulk density* and *thermal conductivity*, the *sorption capacity*, *room acoustics* and *fire performance* is similar to that of plant fibres.

### Possible mixing proportions of earth building materials

Table 3.1 shows empirical values for possible mixing proportions in accordance with the measure of shrinkage of the earth material. The values given serve as a rough guideline for the manufacture of own building materials. The mixing proportions can vary considerably depending on the type of aggregate.

## 3.1.2 Additives

Additives are materials that influence the hardening characteristics of the earth mixture. They are generally only added in small quantities (mass or volume).

### Powdered clay

The addition of powdered clay enriches the naturally-occurring binding agent in earth. This generally helps to improve the *strength* of the material. When using clays with *three layer clay minerals* (2:1 type clay), the *sorption capacity* of the building materials can also be improved (section 2.3.3). Powdered clay in different colours can be used to influence the *appearance* of coloured plasters and rammed earth constructions. Excessive levels of powdered clay can negatively impact on the *drying shrinkage*.

### Mineral binders: lime – cement

Mineral binders are primarily added to earth mixtures to improve their *strength characteristics* and reduce their *sensitivity to moisture*.

Table 3.1: Possible mixing proportions of earth building materials

| BUILDING MATERIAL | Measure of shrinkage, naturally-moist (%) | | | | | | | | | | Measure of shrinkage, malleable mixture (%) | | | | | | | | | | Parts by volume of aggregate to 1 part earth mixture mineral grain aggregate / plant fibre additive (where applicable) |
|---|---|---|---|---|---|---|---|---|---|---|---|---|---|---|---|---|---|---|---|---|---|
| | 1 | 1.5 | 2 | 3 | 4 | 5 | 6 | 7 | 8 | 9 | 10 | | | | | | | | | | |
| | 1.5 | 2.0 | 2.5 | 4 | 5.5 | 7 | 9 | | not testable | | | | | | | | | | | | |
| Earth plaster mortar (t ≤ 1 cm) | 0–0.5 / 0.1 | → | 1 / 0.1 | → | → | 2 / 0.1 | → | → | → | 3 / 0.1 | → |
| Earth plaster mortar (t ≤ 2 cm)* | 0–0.5 / 0.1 | 1 / 0.1 | → | → | 2 / 0.1 | → | 3 / 0.1 | → | → | → | |
| Light earth masonry mortar | – / 0.3 | → | → | – / 0.4 | → | 0.5 / 0.5 | → | → | → | 1 / 1 | → |
| Straw-clay | | | | | | | | | | | |
| Light earth | | | | | | | | | | | |
| Weller | – / 0.3 | → | → | – / 0.4 | | | not used | | | | |
| Rammed earth | 0–0.5 / 0.1 | → | 1 / 0.1 | → | → | 2 / 0.1 | | | not used | | |

* also earth masonry mortar (bulk density > 1200 kg/m³)

Unlike clay, lime and cement harden by chemically transforming the material. This hardening process cannot be reversed through the addition of water. Earth building materials whose water solubility or strength has been modified through the addition of additives or chemical substances are known as *stabilised earth building materials*. In the correct proportions the binding effect of the clay minerals and the added binder can have a cumulative effect. If this is not properly observed, the earth serves as a (not particularly favourable) aggregate. Accordingly, one could term stabilised earth building materials as lime- or cement-bound building materials with earthen impurities. Lime- and cement-stabilised earth building materials are not commonly used in Germany and as such are not described in further detail in this book.

Organic binders: dung – cellulose (adhesive) – starch

In water and other liquids, plant fibres form a paste, for example as they do in the digestive system of animals. The formation process can be accelerated through heat and chemical processes. Such pastes consist of natural long-chain polymers. At a microscopic scale they serve to improve *strength* and decrease *moisture sensitivity* much in the same way as fibres do at a macroscopic scale.

Historically, cow dung has been a common additive. It improves the *workability* and *strength* of the material. Dung-stabilised earth building materials can also be surprisingly *water resistant*.

Cellulose glue and starch are specially-made additional binders. Building materials that contain these additives are therefore also classed as stabilised earth building materials. Their application is widespread, especially in coloured earth plasters. Clay paints and coatings always contain these binder additives. Accordingly, these materials are dealt with in this book.

## 3.2 Mixing

Mixing involves the even distribution of all the components within the mass of the material. This can take place with or without the addition of water. Mixtures made on site are typically mixed with agricultural hand tools or simple mixing machines. For the industrial production of earth building materials, complex mixing plants are used which are either commercially available or have been developed specially for the production of earth building materials. The resulting mixtures can be used as loose building materials (e.g. for rammed earth or earth mortars), be directly processed into building elements or serve as the raw mass for formed building materials such as earth blocks or building boards.

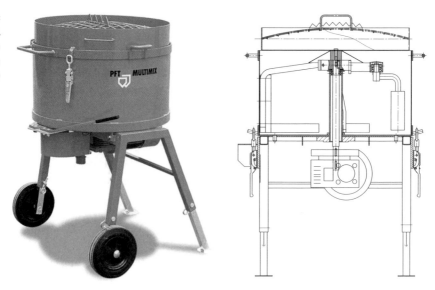

Fig. 3.2 :
Paddle / batch mixer,
PFT MULTIMIX, section
(photo and drawing
courtesy of PFT)

For the manufacture of light earth, earth mixtures are mixed manually or in suitable mixing machines with water to the consistency of a liquid or viscous slurry and then mixed with lightweight aggregates.

Masonry or plaster mortar is made on site from a suitable soil and additional aggregates. The earth material is supplied either in dry powder form or as naturally or mechanically processed material. Depending on how free-flowing the mixture is, a rotary free-fall mixer can be used, but in most cases mortars are mixed with more powerful paddle mixers of the kind used for naturally-moist mixtures (Fig. 3.2).

## 3.3 Properties

Table 3.2 shows the spectrum of the most important properties of earth building materials. The following section details the individual properties and compares them with the properties of other commonly used building materials [Goris, 2008].

### Bulk density

Light earth mixtures are supplied with a raw density of 300 kg/m$^3$ and above. For wet earth construction, bulk densities in excess of 600 kg/m$^3$ are required to ensure sufficient stability of the mass. The bulk density of light earth is therefore of the order of magnitude of softwood. Earth blocks have a bulk density of ≥ 700 kg/m$^3$, earth mortars a bulk density of 1400 - 1800 kg/m$^3$. For rammed earth, the bulk density begins by 1800 kg/m$^3$ and can reach 2400 kg/m$^3$, a density equivalent to that of concrete.

## Porosity

According to tests, the porosity of weller and straw-clay lies between 20 and 30 % [Figgemeier, 1994], [Ziegert, 2002]. For historical earth blocks, a value of 27 % has been determined [Figgemeier, 1994]. These earth building materials therefore have a comparable porosity to other porous masonry building materials. The spectrum can also be used for typical rammed earth, earth mortar and earth blocks. High-density earth blocks can have a lower degree of porosity.

## Drying shrinkage

Earth building materials shrink comparatively significantly as they dry out. In the case of rammed earth and earth mortars, the measure of shrinkage can be 0.5 %. Earth plaster mortars exhibit the largest degree of shrinkage, shrinking by up to 2.5 %. By comparison the measure of shrinkage of concrete or cement mortar is only 0.04 % or 0.09 % respectively. As blocks are dried before installation in a building and shrink as a whole without forming cracks, it is possible to make earth blocks out of richer earth mixtures with a high proportion of sorptive three layer clay minerals. The expected degree of

Table 3.2 : The spectrum of the mechanical, physical and biological properties of earth building materials

| Property | Unit | | from possible | from typical | to typical | to possible |
|---|---|---|---|---|---|---|
| Bulk density | $\rho$ | kg/m³ | 300 | 600 | 2000 | 2400 |
| Porosity | $\varepsilon$ | Vol. % | 20 | 25 | 45 | 55 |
| Measure of shrinkage | | % | 0.5 | 1.0 | 2.5 | 5 |
| Compressive strength | $\sigma$ | N/mm² | 0.6 | 1.0 | 3.0 | 12.0 |
| Modulus of elasticity | E | N/mm² | 300 | 450 | 3000 | 5000 |
| Adhesive strength | $\sigma$ | N/mm² | 0.03 | 0.05 | 0.15 | 0.25 |
| Thermal conductivity | $\lambda$ | W/mK | > 0.1 | 0.17 | 1.10 | 1.4 |
| Thermal capacity | c | J/kgK | | 1000 | 1500 | 1770 |
| Water vapour diffusion resistance factor | $\mu$ | | | 5/10 | | |
| Water absorption coefficient | w | kg/m²h$^{0.5}$ | 1.2 | 6 | 13.4 | |
| Dynamic moisture sorption after 12 hours (depth = 1.5 cm) | | g/m² | 30 | 50 | 70 | 130 |
| Building material class | | | B2 | B1 | A | |

shrinkage can be compensated for by dimensioning the mould slightly larger, a practice already widespread in the brick industry to compensate for firing shrinkage.

Compressive strength

Weller earth mixtures, with a compressive strength of 0.6 N/mm², are the least pressure-resistant earth building materials used for wall construction. A value of 1-3 N/mm³ is more typical for rammed earth and mortars. Non-hydraulic (air-hardening) lime plaster mortars have a similarly low compressive strength (1-1.5 N/mm²) while gypsum plaster mortars have a strength of approximately 3 N/mm². This demonstrates that while not particularly hard, earth plasters are not (as often claimed) remarkably unstable. They exhibit values in the range of other common plaster mortars. Earth blocks have a compressive strength of 2-4 N/mm², at the lower end of the scale given in DIN 1053 for ceramic bricks. In some cases, earth blocks with a compressive strength of up to 12 N/mm² have been measured. By comparison, aerated concrete has a compressive strength of 2.5-10 N/mm², sand-lime bricks 12-20 N/mm² and concrete 30-45 N/mm².

Modulus of elasticity

For earth building materials, the modulus of elasticity typically lies in the range of 450-3000 N/mm². Other building materials exhibit far greater values. For example aerated concrete $E$ 2500 N/mm² (calculated value), masonry brick $E$ 3500 N/mm² (calculated value), concrete C25/30 $E_{cm}$ 26700 N/mm² (secant modulus), steel $E$ 210000 N/mm².

Adhesive strength

Earth plasters should exhibit an adhesive force between the subsurface and the individual plaster layers of at least 0.03 N/mm² [DVL, 2008]. Values of up to 0.15 N/mm² are typical. The adhesive strength of other typical mortars are in most cases higher: lime plaster 0.1-0.2 N/mm², lime-cement plaster 0.2-0.4 N/mm², gypsum plaster 0.4-0.9 N/mm², cement plaster 1.0-2.0 N/mm² [Dettmering/Kollmann, 2001].

Thermal conductivity

The lowest value for the thermal conductivity of light earth lies just above the 0.1 W/mK upper limit for the classification as an insulation material. Typically, earth building materials have a table value of 0.17 W/mK or more (depending on bulk density). Typical insulation materials have values in the region of 0.03-0.06 W/mK, wood 0.13-0.20 W/mK. Up to a bulk density of 2000 kg/m³ the thermal conductivity of earth building materials rises to 1.1 W/mK, which makes them comparable to other masonry building materials with similar bulk densities. In the case of heavy rammed earth, the thermal conductivity can measure up to 1.4 W/mK. By comparison reinforced concrete with a bulk density of 2300 kg/m³ and 1 % steel reinforcement exhibits a much greater thermal conductivity in the region of 2.3 W/mK.

### Thermal capacity

The value for the specific heat capacity of earth building materials ranges from 1000 - 1500 J/kgK depending on composition and bulk density. Organic materials can retain more heat energy than mineral materials. A key influencing factor for the thermal capacity is the bulk density. Most inorganic building and insulation materials have a thermal capacity of approximately 1000 J/kgK. Organic insulation materials such as expanded plastic and cellulose insulation materials have a thermal retention capacity of 1400 - 1600 J/kgK, wood and wood-based materials a thermal retention capacity of 2100 J/kgK.

### Water vapour diffusion resistance factor

DIN 4108 states a single value pair of 5/10 for the vapour diffusion resistance factor of earth building materials. In 2009, a series of tests at the *BAM Federal Institute for Materials Research and Testing* ascertained values for wet measuring of between 8 an 12 (mean value 10) and between 12 and 20 (mean value 17) for dry measurement. At 10/17 the mean values for these earth plasters are clearly much higher than the general values given in the DIN. Overall, however, the tests confirm that earth building materials are vapour permeable. The value 5/10 is also given in the DIN for perforated bricks and aerated concrete blocks and for lime plaster as 5/35, although the upper value is much too high for a pure lime plaster. The values for fired solid and perforated bricks are given as 50/100 and for concrete 60/100 - 80/130 depending on the bulk density. Vapour retarding foils have a vapour diffusion resistance factor of between 1000 and 50,000.

### Water absorption coefficient

The minimum value of the water absorption coefficient of earth building materials is given as 1.2 kg/m²h$^{0.5}$, the mid value in the range 6.0 - 13.4 kg/m²h$^{0.5}$. Earth building materials therefore have a very high degree of capillary conductivity.

For lime and gypsum plasters, the water absorption coefficient is 2.4 kg/m²h$^{0.5}$, for aerated concrete 2 - 6 kg/m²h$^{0.5}$, for lime-sand bricks 3 - 8 kg/m²h$^{0.5}$ and for solid bricks with a bulk density of 1800 kg/m³, 12 - 30 kg/m²h$^{0.5}$. Calcium silicate boards, which are generally regarded as exhibiting high conductivity have a water absorption coefficient of 5 kg/m²h$^{0.5}$. Wood has a water absorption coefficient of just 2 kg/m²h$^{0.5}$ parallel to the grain (end-grain wood), perpendicular to the grain only 0.05 kg/m²h$^{0.5}$.

Dynamic moisture sorption

The dynamic moisture sorption is a measure that describes the water vapour absorption capacity of layers of a building element. To determine this value, a 1.5 cm thick test specimen with equilibrium moisture content (at 50 % relative humidity) is exposed to a relative humidity of 80 %. The values below describe the increase in mass in grams per m² after 12 hours. Earth plaster mortars which contain a low proportion of sorptive three layer clay minerals or are generally very lean exhibit a value of approx. 30 g/m² which equates to the moisture sorption of a lime mortar. The mean value of the dynamic moisture sorption lies between 50 and 70 g/m². Earth plasters therefore exhibit unusually high sorption characteristics. The same applies for untreated wood (60 g/m²). Tempered plasterboard exhibits values of between 6 and 10 g/m². Open-pore paints, for example casein-based coatings have a limiting effect on the sorption characteristics of earth plasters. Tests with wallpapers and emulsion paint on lime-gypsum plasters and plasterboard panels produced values of 15 - 18 g/m². Earth blocks with a high proportion of three layer clay minerals absorb up to 130 g/m². By comparison concrete has a sorption capacity of approx. 30 g/m².

Building material class

Earth is classified in DIN 4102 as *non-flammable* (A). The same applies for mineral aggregates. In the DIN 18951 from 1951 (withdrawn in 1971) earth building materials, including those with appropriate plant fibre additives, with a bulk density of ≤ 1700 kg/m³ are classified as non-flammable. Normed tests showed that straw-clay with a bulk density of > 1200 kg/m³ also fulfils the criteria for non-flammability. By contrast, woodchip-light earth with a bulk density of > 800 kg/m³ is only *flame resistant* (B1) and wood light earth with a typical bulk density of 600 kg/m³ exhibits *normal flammability* (B2). Wood fibre insulation materials likewise exhibit *normal flammability*. Reed panels are also classified as having *normal flammability*.

In terms of the *fire-resistance rating* of earth building materials, the relevant property of earth is the temperature at which chemically-bound crystallisation water is released in the event of fire. This requires energy and as a result retards the heating up of the cold side of the building element. Gypsum also exhibits this quality, although with gypsum the expulsion of moisture bound in the material leads to a loss of strength.

# 4 Earth plasters

## 4.1 The application of earth plasters

### 4.1.1 Reasons for using earth plasters

Earth plasters are chosen over other plasters for their appearance and their contribution to improving the quality of living in interiors. As natural building materials their product life cycle is especially economical with resources. Last but not least, earth plasters are used in the conservation of historical buildings.

The visual and haptic qualities of earth plasters are a product of the colour and granularity of the material, the surface treatment used and the contribution of special additives such as plant fibres (figure 4.1).

Factors that contribute to a better interior climate include the breathability of the surface and the material's pronounced capacity to absorb and give off moisture. Its ability to bind or neutralise odours is a further reason that has as yet not been conclusively proven. Tests at the *Federal Institute for Materials Research and Testing (BAM)* conducted in 2007 found clear indications in this respect (figure 4.2 as well as tables 4.1, 4.2 and 4.3) [Müller, 2007].

The material life cycle of the product from beginning to end and the absence of any synthetically produced and toxic components characterise earth plasters as natural building materials.

In the conservation of historical buildings, earth plasters complement the existing building fabric, not least in technical terms due to their straightforward ability to bond with existing earth plasters. A further aspect is the water-soluble reversibility of its binding qualities which means it can be removed at a later date without damaging the underlying surface.

Fig. 4.1:
Earth plaster in its natural state

## 4.1.2 Application areas of earth plasters

Earth plasters are used in living areas and similar kinds of interiors. It is important, however, to take into account their susceptibility to moisture as well as their low mechanical stability, which is lower than that of lime-cement plasters. Although also somewhat lower than that of lime or gypsum plasters it lies in a similar range.

If earth plasters are to be used on surfaces subject to high wear and abrasion, such as stairwells or corridors, it is necessary to assess their suitability in each specific case.

Earth plasters are generally suitable for rooms which are only temporarily subject to high levels of humidity such as kitchens and bathrooms, unless these are classified as wet rooms in the building regulations. For wet rooms subject to continuously high levels of humidity, such as swimming baths or industrial kitchens, earth plasters are not suitable. In such cases surfaces are required to be easy to clean.

Similarly for surfaces that need to be resistant to chemical attack or other special requirements, earth plasters as well as lime and gypsum plasters are not suitable. Such surfaces require special mortars. This applies especially for underlying surfaces that are contaminated with salts.

Earth plasters can be used in cellars provided the walls are dry and the temperature and relative humidity is similar to that of other living areas.

In outdoor areas, earth plasters can generally only be used on surfaces protected against the effects of weather. That said, the weatherproofing can be improved with the help of good mixtures, special additives and protective coatings. The degree and kind of weathering should be assessed in each case. The experience of contractors offering external earth rendering on surfaces exposed to the weather can be checked by examining their reference projects.

When plastering chimney breasts, stoves or other elements subject to considerable heat, fire safety regulations may require the use of *non-flammable class A building materials*. According to DIN 4102 *Fire behaviour of building materials and elements*, mineral earth plasters fulfil this condition, although fibre-reinforced plasters generally resist cracking better when exposed to heat. Tests have shown that earth plaster mixtures with typical fibrous additions also perform as non-flammable building materials of fire safety class A (section 4.9.4) [DVL, 2009].

### 4.1.3 Aspects of long-term use

Earth plasters can reach a considerable age. In Japan, where the use of lime wash was relatively uncommon, earth plasters have survived for centuries.

Coloured earth plasters are used in the interior design of spaces. As with other internal room finishes, the expectation is that these will survive for decades without the need for appreciable maintenance. Exposed plaster surfaces have to be handled with care, avoiding damage to the surface where possible. When hanging pictures or other objects on the wall, holes should be carefully pre-drilled without using hammer action.

Due to the water-solubility of earth plasters, repairs are generally simple to undertake. It is much easier to repair untreated surfaces than those that have been treated with a stabiliser.

It is advisable to set aside sufficient material for later repairs, especially in the case of coloured earth plasters. On smoothed surfaces only the damaged sections of the surface are smoothed over with a putty knife. A good example for the easy repair of earth plasters is the *Kolumba Museum* in Cologne, whose walls are regularly smoothed over with each new temporary exhibition so that the pictures can be re-hung directly on the wall (figure 4.2).

For rubbed surface finishes, the reworked sections of wall are sponged down with wide sweeping movements after the application of a plaster or putty coat. The same technique can be used to refresh the colour of an entire surface, after gently pre-wetting the surface. When sponging, only a small amount of water is used and the sponge surface, as well as all other tools, must be kept very clean.

For earth plasters that have been painted, repairs can be undertaken using filler with reinforcing fibres or similar. If the visual appearance and feel of the earth plaster surface is no longer desired, it can simply be wallpapered over or painted (section 4.6).

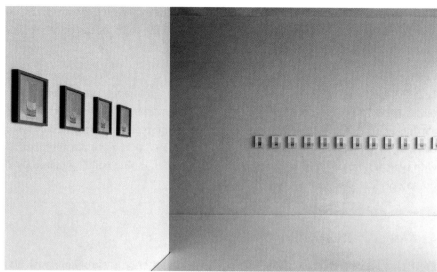

Fig. 4.2: Earth plaster surfaces in the Kolumba Museum in Cologne

## 4.2 Constitution of earth plaster mortars

### 4.2.1 General composition of earth plaster mortars

Earth plasters consist of earth or clay and additives. Additives need to be added to rich, strongly cohesive earth mixtures to reduce the risk of shrinkage cracking. Additives can be organic or mineral (section 3.1.1). In addition to leaning down a mixture they also serve as reinforcement fibres or to improve the thermal insulation properties. A good quality plaster exhibits good cohesion and good mechanical stability along with minimal susceptibility to shrinkage cracking.

Earth mortars can be manufactured from naturally occurring earth from pits, powdered earth or powdered clay in concentrated form. As a rule, all plaster mixtures must be sufficiently cohesive. In addition the mineral structure of the mixture is decisive for the quality of the plaster. Both the kind and form of the granular aggregate as well as the mixing proportions are key factors: Sufficiently coarse-grain mixtures generally have better properties than fine-grain mixtures, especially for thicker coats.

Naturally occurring earth extracted from pits may in some cases have a grain structure that is already fundamentally suitable. Such soils are easy to lean down for manufacturing purposes or mixing on site. When manufacturing earth plasters from powdered clay, however, the entire mineral skeleton has to added in the form of one or more different kinds of sands. Mixtures also need a sufficient amount of fines.

Plant fibres are added primarily to strengthen the inner cohesion of the mixture. As reinforcement they are able to resist stresses in the material resulting from movement in the substrate, shrinkage in thick applications of plaster and other stresses. In addition they act as a compressive buffer accommodating thermal expansion and contraction. Mixtures without plant fibres generally have a higher sand content and therefore less good mechanical properties. The sorption capacity can also be reduced.

Stabilised earth plaster mortars contain other mineral or organic binders in addition to clay, for example lime, cement, gypsum, modified starch or methyl cellulose. Depending on the additive, the water solubility of earth plasters can be lessened or even cancelled.

The quality of earth plaster mortars depends on the homogenous and intimate mixture of the individual constituents.

### 4.2.2 Kinds of earth plaster mortars

Earth plaster mortars can be manufactured on the building site or sourced in premixed form. *Site-made mortars* are mixed out of individual constituents on the building site. *Semi-finished factory-made mortars* consist of prebatched binding mixtures which are mixed on the building site with sand according to the manufacturer's instructions. *Factory-made mortars* are finished products that only need mixing with water on site.

*Naturally-moist earth plasters* have a similar degree of moisture to earth as excavated out of the ground. The specific hardening behaviour of earth makes it possible to store earth plaster mortars in a moist consistency after mixing. This delivery form is both economical and ecological because no energy is required to dry the mass. Naturally-moist earth plasters are delivered in lightweight textile containers called "big-bags". These 1.5 t heavy bags can necessitate particular on-site logistics. Naturally-moist earth mortars can only be processed by particular plastering machines; typical gypsum plastering machines designed for use with dry material are generally not suitable (section 4.5.1). It is important to note that mortar that has frozen over winter cannot be used until it has fully thawed.

The proportion of microbiological germs in naturally-moist mixtures cannot realistically be restricted according to kind or quantity to 'guaranteed' levels (section 3.1.1). Nevertheless, simple quality assurance measures, such as storing in a dry place and visual monitoring of plant additives, are possible. The shelf-life of naturally-moist mixtures is limited: after a few months the straw will begin to humify and can no longer lean down or reinforce the mixture.

*Dry earth plasters* consist of a series of parent substances that are dried separately and then mixed together. Another commonly used approach is to mix in a naturally-moist state and then to dry the entire mixture.

Dry earth mortars are packaged in big bags, paper sacks or silos. At present only earth mixtures without fibrous additives are delivered in silos.

Dry earth plasters can be worked with normal gypsum plastering machines (section 4.5.1). This is very often the most important reason for choosing this packaging form.

### 4.2.3 Grades of earth plaster mortar

Types of earth plaster are differentiated according to their function and granularity.

*Undercoat plasters* have a composition that is not conceived for use as finishing plaster layers. They typically contain relative coarse constituents and are rich mixtures. The high proportion of binding agents they contain improves the adhesion but is also more susceptible to crack formation.

*Topcoat plasters* or *final coat plasters* are conceived for the facing layer and exhibit minimal crack formation when applied in the recommended maximum thickness.

*Universal plasters* are not clearly defined. The term can refer both to universal applicability on different substrates and for different purposes as well as their suitability for use as an undercoat and topcoat plaster.

*Fine-finish plasters* are topcoat plasters applied in a thin skim coat with a very fine granularity of < 1 mm. They are also used as undercoat plasters for further thin layer coats of plasters.

*Smoothing plasters* have an especially fine granularity suitable for smoothing coats.

*Coloured plasters* are selected and applied in the desired final colour and not painted over. As a result they should exhibit good abrasion resistance. Very often, they are stabilised with cellulose or starch additives in addition to the clay minerals. Their colour results either from the use of coloured clays or sands or by colouring with pigments.

*Brush-on plasters* are not plasters in the true sense of the word but textured paints. More properly they could be termed *slurries* or *washes*. However, due to their plaster-like surface qualities, they have become known as brush-on plasters. *Earth paints* do not contain any visibly granular particles. Brush-on plasters and earth paints always contain a further binding agent alongside clay.and are described separately in chapter 5.

## 4.3 Substrates for earth plasters

### 4.3.1 Plaster substrates in general

As for all paints and plasters, substrates for earth plasters need to be *firm, rough, sufficiently absorbent* and *dry*. They should *not be soiled, dusty* or *exhibit stains* that could show through, nor should they be contaminated with *salts* or contain *frost*. Depending on the uniformity of finish desired, they should be sufficiently *smooth* and *level*. For plaster applications on ceilings, the required mechanical properties of the substrate are greater than that of the walls.

The plasterer assesses the mechanical suitability of the substrate and other aspects based on a visual inspection and his knowledge of the possibilities of his trade. It can be most informative to conduct a trial on a sufficiently large surface. Different earth mortars can exhibit quite different tolerances with regard to the quality of the substrate. According to the German Construction Contract Procedures, VOB Part C DIN 18350, the plasterer should raise concerns where:

- the condition of the substrate is unsuitable, for example it exhibits clear surface contaminants, efflorescence (e.g. visible salt contamination), is too smooth, exhibits oily patches, uneven suction, frozen sections or a variety of different kinds of surface materials
- the building site or materials has high moisture or humidity levels
- surfaces are more uneven than permitted according to DIN 18202
- there are insufficient anchoring possibilities
- there are no datum lines on each floor.

The required *strength* of the substrate for earth plasters is no greater than for other plaster mortars. As earth plasters form a comparatively soft and low-stress layer, they can also be applied to very soft mineral substrates. Loose or flaky stone or plaster remains must be removed. Loose means that they can be removed by hand without great effort. Sandy constituents should be brushed off; where old plasters are sandy, they may need stabilising. Coats of paint can only remain on the substrate if they are absolutely stable. If the substrate consists of building boards, these must be sufficiently level and arranged in bond. They should not give under pressure: pressing with one's thumb provides sufficient indication of stability.

Sufficient surface *roughness* or *texture* is a comparatively important criterion as earth mortars adhere mechanically. Thick applications and applications subject to specific stresses, such as those containing embedded wall heating, adhere better on substrates with rougher surfaces. The method of application can also have a decisive effect: a preparatory spatter coat with earth mortar can improve adhesion significantly.

For the mortar to adhere properly, *sufficient absorbency* – i.e. good suction characteristics – is likewise important. A lack of surface absorbency can be compensated for with surface roughness to a limited degree and vice versa. For this reason, granular and highly adhesive primers are used to prepare some low-suction surfaces for plastering.

Only sufficiently *dry* substrates will be absorbent. If the pores are filled with water, the surface will not absorb any more moisture. This is why when pre-wetting surfaces they should not be saturated with water but wetted with a fine spray mist. Particular attention should be paid to massive wall constructions that have been exposed to rain for a prolonged period where the pores are partially or fully saturated with water. In most cases, visual assessment is sufficient to establish if previous wet earth constructions (rammed earth, light earth) are dry enough to plaster over but if in doubt, the moisture content can be ascertained more precisely by oven-drying a test sample and comparing the relative weight before and after. In addition to the aforementioned mechanical aspects, wet substrates can hinder the drying out of earth plasters considerably and possibly lead to damages (section 4.5.6).

Substrates that are not absorbent but sufficiently rough such as wood-wool insulation board or reed mat insulation boards should not be wetted as the water then acts as a separating film.

Surfaces that are not rough or not absorbent can lead to increased crack formation in the earth plaster. In such cases the lack of adhesion to the substrate means that the background surface does not help in resisting plaster shrinkage. Where the plaster otherwise forms many small and insignificant shrinkage cracks, here shrinkage results in fewer but larger cracks.

The requirement that substrates be *free of contaminants* (for example shuttering release agents) and *dust* is no different to when using other plaster mortars. Substrates with tar, nicotine or soot *stains*, for example around chimney breasts, must either be replaced or contained using the painter's usual tools of the trade, or else they will show through the plaster.

Particular attention should be paid to possible *salt contamination* in the substrate. This applies not least because earth mortars are widely used in the rehabilitation of existing buildings. Here it is worth noting that the moisture sorption characteristics of earth plasters are sometimes confused with the properties of special *restoration plasters*. Soluble salts transported with moisture in masonry damage mortar by expanding in volume as they crystallise. Restoration plasters are designed to have a coarse pore structure that provides sufficient space for salt crystals to expand within it. They are hydrophobic and have a rigid, cement-based mineral binding matrix. Earth plasters, like lime and gypsum plasters, do not have these qualities. Their pore structure is not as open and the binding effect of the clay minerals comparatively weak. For these reasons, earth plasters are even used as so-called "sacrificial plaster layers" with the aim of desalinating a wall: after they have absorbed soluble salts, they can easily be removed.

Salt contamination is often not visible with the naked eye and when in doubt an analysis must be conducted. Common situations include old brickwork walls that stand directly on moist ground as well as buildings formerly used for keeping livestock. Sometimes the fact that the base of a wall has obviously needed repeated repairs in the past or has been given a bituminous coating is a sign of possible salt contamination.

Earth plasters are, however, suitable as a topcoat for restoration plasters due to their high vapour permeability.

### 4.3.2 Common plaster substrates

The following section provides an overview of the key aspects to consider when applying earth plaster to different substrates. These must be supplemented by a thorough and expert examination on site in each and every specific case. Similarly, it is absolutely necessary to adhere to the manufacturer's recommendations, particularly given the fact that the products available on the market vary considerably in their properties and the surface preparation involved. When in doubt, trials should be conducted on sufficiently large surfaces.

*Earth blocks* are well-suited for receiving earth plaster as long as they correspond to Usage Class II or better. They must be sufficiently stable and should not swell appreciably when exposed to water (section 8.2) [DVL, 2009]. Earth blocks with very smooth surfaces may need to be prepared through the application of a spatter coat.

When plastering *clay plasterboard*, the manufacturer's recommendations must be followed. The application of reinforcement scrim tape to the joints is always necessary. Only very thin layers should be applied to extruded boards, as these are more susceptible to swelling and warping when exposed to water.

*Rammed earth* must be sufficiently dry before plastering and the possibility of further crack formation, as well as wall settlement, should be eliminated before plastering as any later cracking or movement in the substrate will show through to the plaster surface.

*Light earth* constructions are generally held in place with permanent formwork that forms a good plaster base for plastering with earth, for example reed plaster lath. Open formwork made of slender timber boards must first be faced with a plaster lath. Where light earth surfaces are to be plastered directly (e.g. where temporary shuttering has been employed), the surface must be sufficiently stable to hold the plaster. The light earth mixture must always be fully dry before applying earth plaster.

Brick masonry made with *solid bricks* is generally a straightforward plaster substrate. In the case of *porous bricks* and similar products, the suction characteristics should first be assessed. Many products only start exhibiting strong suction after they have been wetted once. The wetting of *perforated bricks with vertical perforations* for better insulation should be undertaken with care as the narrow wall thickness of the bricks can quickly become saturated with water, causing them to lose their suction characteristic. *Lime-sand brickwork* with thin mortar joints can present a very smooth surface that may need pre-treating much like a concrete wall (see below). In many cases a light spatter coat with earth mortar may be sufficient to provide the necessary adhesion.

*Concrete* is a difficult plaster substrate. The surface must be cleaned of all shuttering release oils, cement paste and free of residual moisture. This can be quickly ascertained by *test wetting* the surface: water is applied to the surface with a brush and should evaporate or soak away within a matter of minutes. If a surface remains moist, the lack of absorbency may be attributable to one of the above reasons. Oils can be washed off with suitable cleaning agents, layers of cement paste must be fluate treated, while residual moisture must simply be allowed to dry. Smooth concrete surfaces are given a spatter coat of cement and coarse sand or are alternatively treated with a suitable primer, which is now more common. Even with very good preparation, layers applied to concrete ceilings should not be too thick. It is highly recommended to conduct trial surfaces in advance of plastering.

Masonry made with *aerated concrete* has a very pronounced absorbency. To increase the working time of the plaster and to minimise shrinkage cracking in thin layers of earth plaster, the absorbency can be reduced with a suitable priming agent. An alternative approach is to apply multiple coats of plaster.

*Old earth plasters* in existing buildings need good preparation despite the material compatibility. To begin with loose earth plaster must be removed, the surface then carefully brushed down and wetted with a spray mist of water so that the strong connection between old and new earth plaster is not impaired by dust or sand. In a next stage the old earth plaster is rubbed down while wet with a hard brush or a sponged or felted rubbing board. Even better is the working in of a thin layer of new coarse earth plaster in the same process. This is essential if only a thin layer of topcoat plaster is to be applied to the old plaster. Careful preparation is particularly important where the existing plaster is very lean and contains little straw. The properties of the new earth plaster should as far as possible correspond to those of the existing plaster. The preparations described here are just as necessary for filling holes in the existing plaster as they are for applying a new surface coat. In most cases two layers are applied to existing earth plaster as very often timber members and installation chases need to be covered over or uneven patches levelled out. A coarse earth mortar should be used for the base coat. Special attention should be given to the possibility of salt contamination at the base of the wall – this may still be the case even if the existing earth plaster appears to be intact. New uses and heating patterns can alter the previous prevailing pattern of moisture production, causing salt-induced damages to appear where they were not previously visible.

*New earth plaster* can only be plastered over once they are dry enough that no further shrinkage cracks will arise. Cracks that occur later in the base coat show through on the topcoat and are difficult to repair. If a coloured fine-finish plaster is to be applied, the undercoat plaster has to be free of cracks and level. If necessary small cracks can be wiped closed, large cracks filled with a new thin filler coat. To achieve an absolutely flat, fine-grain finish, it can be beneficial to first apply a thin preparatory coat of fine-finish earth plaster, if desired already in the respective colour.

Loose sections of *old paint* should be removed. Stable paint usually only needs priming to improve adhesion of the subsequent plaster layer.

*Lime or gypsum plasters* can be plastered over as long as they are sufficiently rough, firm and stable. New mineral plasters may have low absorbency. The absorbency of old mineral plasters, by contrast, can vary significantly and it can be necessary to apply a primer to even out their suction characteristics.

*Gypsum plasterboard* or *gypsum fibreboard* are not suitable for receiving thicker layers of plaster. If they become overlay wet they run the risk of deforming causing damage. In addition plasterboard is not designed to support further surface loads. Thick layers of plaster should not be applied. Thin layers of up to 3 mm thick are generally possible when the substrate has been well-prepared. To begin with the board joints must be filled and reinforced with mesh scrim; a fibrous reinforcing putty is on its own not sufficient. Applying self-adhesive joint tape before applying the filler putty is even more secure. Boards without recessed joints need to be reinforced over their entire surface. The reinforcement mesh is applied with a thin bed of filler mortar. After drying, the surface is treated with an appropriate primer. The primer serves to even out the suction characteristics of the panel surfaces and the joints and protects the panels from absorbing too much moisture and deforming. Granular primers also provide a degree of surface roughness.

*Magnesite-bonded wood-wool lightweight building board* is a substrate with an especially textured surface that has been used for years in earth building. The manufacturer presupposes the use of 50 mm thick board (Heraklith BM). They are glued at their joints with a special adhesive. DIN 1102 11-1989 states that "panels that are not directly fixed to a massive subsurface and are intended to be plastered, for example in timber stud constructions, roof inclines, batten subconstructions and similar..." should be stabilised with a spatter coat of mortar of the mortar class P III. In contrast to this, it has become common practice to plaster them directly with earth plaster as long the as construction – the panel thickness and its ubconstruction – is sufficiently stable.

*Wood fibre insulation board* can be plastered with earth plaster as long as it is sufficiently stable and textured. The texture can be improved by moderate roughening of the surface. The boards are usually plastered with two coats of 2 - 3 mm thick layers of fine-finish plaster. If thicker layers of up to 15 mm in total are to be applied, the surface should be primed in advance. For even thicker layers, for example for embedded wall heating, the necessary stability and mechanical key can be achieved by applying a bonding layer applied with a notched spatula. In general, it is important that individual layers are not applied too thickly to avoid exposing the boards to prolonged moisture. In addition the drying of thick plaster layers is problematic on substrates with little or no suction (section 4.5.6).

*Wood planking* must be clad with plaster lath before it can be plastered. The planking boards should be set slightly apart to avoid it skewing as a result of swelling deformations after the plaster has been applied. Squared timbers should likewise be clad with plaster lath. When using fibre-reinforced mortar, they can be plastered directly up to a width of 6 cm (see manufacturer's instructions!).

Hard composite wood boards such as *OSB* or *chipboard* are not suitable for direct plastering as even a small amount of moisture causes significant deformations.

Plaster lath boards such as *reed* can be plastered directly. The same applies for other sufficiently stable *insulation boards*. Smooth and highly absorbent materials such as *calcium silicate insulation board* should be treated with a granular primer to improve surface adhesion and reduce absorbency. The boards are usually used to treat moisture problems and for internal insulation. For this reason a primer coat should be vapour permeable. *Mineral foam insulation boards* should be treated similarly to aerated concrete.

*Straw bales* are typically plastered with a strongly adhesive and comparatively thick undercoat layer of not too stiff earth mortar. The plaster can be spray-applied under high pressure (with a plastering machine) or vigorously worked in manually (hand plastered). Recesses, gaps and cavities in the structure of the straw bales must be well filled. The subsequent layers are then applied as normal.

### 4.3.3 Primers and absorption sealers

Primers are generally casein-, dispersion- or silicate-based with a dispersion proportion. They are used to prime surfaces in preparation for receiving earth plasters to even out irregular suction characteristics or to protect underlying surfaces from absorbing excess moisture. As the binding agent in casein-based primers is susceptible to water, casein primers are usually only applied under thin layers of plaster. Casein-bonded products also need substrates with a higher absorbency than, for example, dispersion primers do. On surfaces with little to no suction, polyurethane or epoxy resin primers have to be used. Textured primers improve the surface roughness.

Some plaster manufacturers prescribe primer coats made of earth or powdered clay slurry, in order to improve adhesion. This is most appropriate for lean plaster mortars and plaster mortars with poor adhesion characteristics. For adhesive plaster mortars, the slurry coat can function as a separating layer and actually impede the adhesion.

Due to the special hardening characteristics of clay-bonded building materials, absorption sealers are generally not required for earth mortars (section 2.3). To increase the working time, in all but a few special cases it is generally sufficient to moderately pre-wet the surface. For earth substrates, this improves the adhesion between the plaster and the subsurface.

### 4.3.4 Plaster laths

Plaster laths provide a largely independent mechanical key for plastering substrates with little own surface adhesion, for example wooden surfaces or mixed mineral substrates. They need to be fixed durably and sufficiently often so that they are not or only very slightly springy. On timber beams, they can be underlaid with paper. Because in interiors wood is generally subject to very little swelling deformation, the paper underlay may not be necessary when using fibre-reinforced mortars (figure 4.3). As a rule, and particularly where mixed mineral substrates are concerned, sufficiently wide-meshed products should be used, appropriate for the planned granularity of the plaster mortar. Only then can the plaster mortar penetrate the mesh and ensure good contact with the substrate. It is imperative that the plaster lath does not cause hollows to form behind the plaster layer, particularly where mineral substrates are concerned. Finally, plaster laths should not be confused with reinforcement mesh. The latter serve to bridge joints between building boards or to resist stresses occurring in the upper third of the plaster thickness (section 4.4.3).

*Reed plaster lath* consists of ca. 6-10 mm thick reed stems that are laid by machine on an approx. 1 mm thick galvanized carrier wire and bound to it with a second much thinner wire. The binding wires are spaced ca. 10-20 cm apart. If the plaster lath is to be stapled on the wire, it is important to staple the stronger carier wire so that the reed lath is pressed firmly against the substrate.

The stems must be spaced sufficiently far apart so that the mortar can easily pass between them. Plaster laths with approx. 70 stems per metre are commonly used.

Fig. 4.3 :
Reed plaster lath applied to the surface of a timber post

*Brick wire mesh* consists of approx. 1 mm thick wire with a mesh distance of 20 × 20 mm. Small pieces of fired clay approx. 7 mm thick are bonded to the nodes where the wires cross. The mesh can be bent in different directions and is therefore often used to plaster around the ends of beams or to form a base for spherical bulging forms.

*Expanded metal plaster lath* is available in a wide variety of different quantities on the market. *Lochrip 0.3* for example has a material thickness of 0.3 mm with a rib height of 10 mm. Expanded metal mesh is good for relatively fine-grain mortars, but the distance between the ribs can be too small for more coarse earth plasters. Coarse expanded metal meshes are, by contrast, very rigid and are too stable for use with earth plasters. Their strong susceptibility to thermal expansion can lead to crack formation.

### 4.3.5 Special considerations for substrates for coloured plasters

Substrates for coloured earth plasters must be *especially smooth, free of cracks, uniformly absorbent* and, in addition, have an *even surface colouring*. The impression of colour is strongly influenced by how light reflects off surfaces. The wish to achieve a uniform colour for a wall is therefore very much dependent on the quality of the surface finish. The smoothness or roughness of a plastered surface is strongly influenced by how wet or dry the surface is when it is rubbed or felted. If individual sections of plaster are wetter than others when rubbing, they will exhibit a rougher texture. When dry these sections scatter light differently causing the colour of the wall to look inconsistent.

It is important that the surface of the substrate is *as flat as possible* so that the application of a coloured plaster coat does not exhibit different thicknesses across its surface, which then dry inconsistently. Thick applications of coloured earth plaster may also exhibit cracking. If necessary, apply a levelling filler coat with fine-grain earth plaster mortar in advance of plastering.

*Cracks* in earth undercoat plasters can show through to the top surface. Small cracks can be sponged closed, larger cracks should be filled with a suitable earth plaster mortar. Depending on the intensity of crack formation, it may be necessary to apply a thin fine-grain earth plaster across the entire surface in advance of applying the topcoat.

A more *uniform absorbency* of the substrate can be achieved using primers that also retard the absorbency characteristics and therefore prolong the working time of the applied plaster. Primers are only able to even out a limited degree of differential absorbency and it is therefore important to use materials that are similar to the existing substrate when filling holes etc. Metal components such as edging or corner beads or plaster profiles must be buried deep enough beneath the plaster. If they lie directly beneath the applied coloured plaster they show through on the surface. Pre-wetting the surface also helps to even out suction and extend the working time of the applied plaster. This should be undertaken carefully with a fine spray so as not to saturate the pores of the underlying material. A further option for evening out the suction characteristics of the substrate is to apply a thin coat of fine-grain earth plaster mortar. Many plasterers favour the application of two coats of coloured earth plaster as a means of providing a uniform substrate. After the first very thin coat has dried, the application and working characteristics of the final coat is much more predictable.

Substrates for coloured plasters should have an *even surface colouring*. Dark-coloured substrates can shine through thin coats of light-coloured plaster. This is especially important to consider when plastering a room with mixed substrates, e.g. where some walls are dark earth plaster and others gypsum plasterboard. The application of a white earth plaster will not appear consistent without first priming the darker surfaces with a light-coloured primer or suitable paint.

Overall, the careful preparation of substrates for coloured earth plasters pays dividends when applying the coloured plaster. Such plasters need to be worked swiftly and evenly and the overall result will be incomparably better.

### 4.3.6 Surface quality levels of plaster and drywall substrates for subsequent plastering

The quality of plaster surfaces that are to be painted or given only a thin topcoat plaster must be defined in specifications and contracts in terms of suitability, flatness and aesthetic requirements. Subjective and unclear designations such as "ready for painting" or "free of undulations when side-lit" can lead to disputes.

The overview of *Surface Quality Levels for Internal Plaster Surfaces* described in DIN V 18550: 2005-04 provides a classification system for the necessary surface quality according to the kind of plastering system and its further treatment, i.e. the subsequent layers to apply. These cover surface flatness requirements according to DIN 18202. The quality levels are abbreviated by the letter Q and a corresponding number. Q1 is the most basic quality level, Q4 the highest quality level.

Plaster surfaces are defined by the way in which they are worked, or more precisely the tools used to work the surface:
- Levelled plaster surfaces (featheredge, plasterer's darby)
- Felted plaster surfaces (felted or sponged float) *or* rubbed plaster surfaces (wooden float)
- Smoothed plaster surfaces (smoothing or polishing trowel)

If no specific details regarding the surface treatment are defined in the specification, the rubbed (felted, rubbed) plaster surface finish is assumed. The quality levels Q2-Q4 must always be stated along with the surface treatment method (levelled, felted or rubbed, smoothed). If no specific quality level is defined in the specification, quality level Q2 (Standard) is assumed. The notes regarding suitability for subsequent possible surface coatings contained in the quality levels should be understood only as examples. The specification should always explicitly state any subsequent surface coverings or treatments; general indications are not sufficient.

If the assessment of the plaster surface or acceptance of the final works by the client will take place under particular lighting conditions, such as natural lighting from the side or artificial lighting, the client should ensure that comparable lighting conditions prevail during the plastering works. Because lighting conditions are rarely constant, it is only possible to clearly assess the surface quality for one single predefined lighting situation. This lighting situation should be contractually agreed in advance. In the case of Q4, the lighting conditions of the room as it will be used should be described in the specification.

With regard to the flatness tolerances (according to DIN 18202) cited in the definition of the quality levels for internal plasters, it should be noted that these tolerances need only apply where sensible and necessary for the end requirements. This means that plasters with inherent dimensional tolerances that do not impair their technical function or the desired appearance (for example at junctions or under particular lighting conditions) are not necessarily insufficient [Franz/Schwarz/Weißert, 2008]. Where very particular surface quality requirements are desired, these should be explicitly defined in the specification and contractually agreed.

## 4.3 SUBSTRATES FOR EARTH PLASTERS

The *Bundesverband der Gipsindustrie e. V. (German Association of the Gypsum Industry)* has published worksheets explaining the quality levels. The sheet *Putzoberflächen im Innenbereich (Plaster Surfaces in Interiors)* details the surface qualities of wet applied plasters. A corresponding publication for drywall plaster surfaces has also been published: *Verspachtelung von Gipsplatten, Oberflächengüten (The Trowelling of Gypsum Fibre Boards, Surface Qualities)* [GIPS, 2003], [GIPS, 2007].

The direct applicability of these requirements for substrates made of earth plaster should be viewed critically, as earth plasters generally contain more coarse mineral and organic constituents than gypsum plasters. It is even more difficult to apply the classification system to describe the desired visual qualities of coarse or coloured plaster surfaces made of earth mortars (section 4.7.4): on the one hand the quality levels describe *the quality of substrates* for subsequent plastering and *not the final surfaces* themselves; on the other hand, aggregates in earth plasters as well as the working method during application are arguably a characteristic quality of earth plasters.

That said, we have nevertheless elected to orient our description of surface qualities on the above classification system. Its use as a means of describing the surface qualities of fine-grain mineral plasters or gypsum fibre boards is well-established, although for many earth plaster applications it cannot be applied without modifications or qualification. Where detailed, the requirements for lime plasters can, for the most part, be applied to earth plasters.

For substrates for earth plasters, the following requirements are proposed:
- Levelled plaster surfaces          Q3
- Felted or rubbed plaster surfaces  Q3
- Smoothed plaster surfaces          Q2
- Trowelled plaster surfaces         Q3

In contrast to the original tables, each plaster and drywall surface has been detailed in an individual table. Here too the quality levels are defined in each column rather than row. The rows describe the categories *requirements, suitability, flatness tolerance, execution* and *limitations*. This arrangement makes it possible to incorporate various preliminary remarks from the work sheets into the tables, which are necessary to know in order to properly apply the quality classification system. Similarly, it has also been possible to integrate the flatness tolerances according to DIN 18202 in the table.

## Levelled plaster surfaces

Table 4.1 : Characteristics of quality levels for internal plaster substrates with levelled plaster surfaces

| | Q1 | Q2 | Q3 | Q4 |
|---|---|---|---|---|
| Requirements | For surfaces with few or no cosmetic or flatness requirements – a closed plaster surface is sufficient (simple, rough plaster application). | For plaster or undercoat plaster surfaces with few or no cosmetic requirements, but with normal flatness requirement (contractually agreed) – a levelled plaster surface is sufficient. | For plaster or undercoat plaster surfaces with few or no cosmetic requirements, but with better flatness requirement (contractually agreed) – a smooth levelled plaster surface is required. | – |
| Suitability | n. a. | – Decorative topcoat plaster ≥ 2.0 mm<br>– Wall coverings made of ceramics, natural or artificial stone, etc. | – Decorative topcoat plaster ≥ 2.0 mm<br>– Wall coverings made of ceramics, large-format tiles, glass, natural or artificial stone, etc. | – |
| Flatness tolerances<br><br>DIN 18202 1997-4 table 3, group 6/7 | n. a. | Distance between measuring points, in m, up to<br><br>0.1   1   4   10   15<br><br>Permitted position deviations, in mm<br><br>3   5   10   20   25 | Distance between measuring points, in m, up to<br><br>0.1   1   4   10   15<br><br>Permitted position deviations, in mm<br><br>2   3   8   15   20 | – |
| Application | n. a. | Single-coat or multi-coat on a possibly prepared substrate. After application the plaster is levelled with a straight edge to achieve a flat surface. Suitable as a base for ceramic or stone tiling or similar. The surface should not be stabilised (e.g. felted or smoothed). | Additional use of plaster beads and profiles to improve surface flatness. Suitable as a base for ceramic or stoneware tiling or similar. The surface should not be stabilised (e.g. felted or smoothed). | – |
| Limits | In single-layer applications, possible shrinkage cracking or slight indentation of the plaster over joints in an uneven substrate are not entirely avoidable. | n. a. | n. a. | – |

## 4.3 SUBSTRATES FOR EARTH PLASTERS

## Felted or rubbed plaster surfaces

Table 4.2 : Characteristics of quality levels for internal plaster substrates with felted or rubbed surfaces

| Q1 | Q2 | Q3 | Q4 | |
|---|---|---|---|---|
| – | Standard quality requirements, sufficient for general requirements for walls and ceilings. | Better quality requirements than Q2 – felted. The textured structure of the felted surface should be even across the respective surface. Granular accumulations or flat spots only permissible in isolated cases and should not impair the overall visual impression of the surface. | Best quality requirements, only achievable with additional measures over and above those of Q3. The textured structure of the felted surface must be absolutely uniform. Granular accumulations or flat spots not permissible. Homogenous overall visual impression. | Requirements |
| – | Felted/rubbed: – matt, non-textured coatings and coverings. Rubbed: – Coarse textured wall coverings, e.g. wood-chip wallpaper, coarse-grain (RG, DIN 6742). | – Matt, non-textured coatings and coverings. | n. a. | Suitability |
| – | Dist. measuring points, in m, up to<br>0.1  1   4   10   15<br>Permitted position deviations, mm<br>3   5   10   20   25 | Dist. measuring points, in m, up to<br>0.1  1   4   10   15<br>Permitted position deviations, mm<br>3   5   10   20   25 | Dist. measuring points, in m, up to<br>0.1  1   4   10   15<br>Permitted position deviations, mm<br>2   3   8   15   20 | Flatness tolerances<br><br>DIN 18202 1997-4 table 3, group 6/7 |
| – | Single-coat felted plaster (gypsum, lime, lime-cement or cement plaster):<br>After application, levelling and alignment of the plaster, the surface is felted with a float.<br>Single-coat rubbed plaster (lime and lime-cement plaster):<br>After application, levelling and alignment of the plaster, the surface is rubbed with a float. | Felted plaster (gypsum-lime, lime-gypsum, lime, lime-cement or cement plaster):<br>After application and levelling, alignment of the plaster.<br>For gypsum-based plasters, plaster surface typically felted before and after.<br>For lime, lime-cement and cement plasters, a second layer of plaster is generally applied and felted. | Undercoat plaster must be at least Q3 levelled plaster with better than normal flatness. For this, plaster profiles, grounds or beads should be used (where necessary removed and replaced after application of undercoat).<br>Q4 is achieved only the through application of an additional layer of decorative felted plaster, when desired with coating/covering. | Application |
| – | For single-coat plasters, occasional shrinkage cracks or slight depressions at joins between different substrates cannot be ruled out.<br>Occasional contours such as flat spots, traces of working, small uneven areas, granular accumulations cannot be entirely ruled out. Slightly different plaster structure despite primer / absorbency sealer not avoidable.<br>Visible undulations when side-lit possible. | Visible undulations when side-lit cannot be entirely ruled out. | Possibility of visible contours minimised, undesirable effects such as undulations when side-lit largely eradicated.<br>Lighting conditions for later use must be known and ideally replicated on site during construction. The technical limitations of working on site should be noted. Plaster surfaces that are absolutely flat and free of undulations when side-lit are not technically not feasible. | Limits |

## Smoothed plaster surfaces

Table 4.3 : Characteristics of quality levels for internal plaster substrates with smoothed surfaces

|  | Q1 | Q2 | Q3 | Q4 |
|---|---|---|---|---|
| Requirements | – | Standard quality requirement, sufficient for general requirements for walls and ceilings. | Better quality requirements, only achievable with additional measures over and above those of Q2. | Best quality requirements, only achievable with additional measures over and above those of Q3. |
| Suitability | – | – Decorative topcoat pl. > 1.0 mm.<br>– Medium to coarse textured wall coverings, e.g. wood-chip wallpaper medium / coarse-grain (RM/RG, DIN 6742).<br>– Matt, textured paintwork and wall coverings, e.g. dispersion paint, applied with sheepswool or textured paint roller. | – Decorative topcoat pl. ≤ 1.0 mm.<br>– Fine-textured wall coverings.<br>– Matt, fine-textured paintwork and wall coverings. | Smooth or textured wall covering with gloss-effect, e.g.:<br>– Metal, vinyl or silk wallpaper.<br>– Varnish or paintwork/wall coverings with semi-gloss effect.<br>– Trowelling / smoothing methods. |
| Flatness tolerances<br><br>DIN 18202 1997-4 table 3, group 6/7 | – | Dist. measuring points, in m, up to<br>0.1  1  4  10  15<br>Permitted position deviations, mm<br>3  5  10  20  25 | Dist. measuring points, in m, up to<br>0.1  1  4  10  15<br>Permitted position deviations, mm<br>3  5  10  20  25 | Dist. measuring points, in m, up to<br>0.1  1  4  10  15<br>Permitted position deviations, mm<br>2  3  8  15  20 |
| Application | – | Single-coat: After application of gypsum plaster or gypsum-based plaster onto existing substrate (prepared where necessary), levelling and alignment of plaster. Additional felting to smooth the surface.<br><br>Two-coat: Suitable smoothing coat applied to an existing substrate (prepared where necessary) of flat, roughly dressed and cured undercoat plaster made of gypsum, gypsum-lime, lime-gypsum, lime or lime-cement plaster. | All works as per Q2. In addition, a subsequent smoothing work stage or application of a smoothing plaster coat. | Plaster must fulfil more stringent flatness requirements. Additional use of plaster beads / profiles or grounds to improve surface flatness (removal of grounds and filling where appropriate).<br><br>All works as per Q3. Additional full reworking of surface with a suitable smoothing coat.<br><br>In specific cases (gloss paintwork, varnishes, shiny wallpaper) further measures may be necessary (repeated filling and sanding) to prepare the surface. |
| Limits | – | Individual traces of working such as trowel marks not entirely avoidable.<br>Visible undulations when side-lit possible. | Traces of working, such as trowel marks, largely eradicated; amount and visibility much less than Q2.<br>Visible undulations when side-lit cannot be entirely ruled out. | Possibility of visible contours minimised, undesirable effects such as undulations when side-lit largely eradicated.<br>Lighting conditions for later use must be known and ideally replicated on site during construction. The technical limitations of working on site should be noted. Plaster surfaces that are absolutely flat and free of undulations when side-lit are not technically not feasible. |

## 4.3 SUBSTRATES FOR EARTH PLASTERS

## Trowelled drywall gypsum panels

Table 4.4 : Characteristics of quality levels for trowelled gypsum plasterboard surfaces (selection)

| Q1 | Q2 | Q3 | Q4 | |
|---|---|---|---|---|
| Basic stopping of joints, sufficient when few or no cosmetic requirements. | Standard trowelling, sufficient for most general requirements for walls and ceilings. | Better quality requirements, additional measures necessary over and above basic / standard trowelling (e.g. special filling). | Best quality requirements. | Requirements |
| n. a. | – Medium and coarse textured wall coverings, e.g. wood-chip wallpaper, medium / coarse-grain (RM / RG, DIN 6742)<br>– Matt, covering paintwork / wall covering (e.g dispersion paint), applied manually or with textured paint roller.<br>– Topcoat plaster, with grain size > 1 mm as permitted by the respective manufacturer of the plasterboard system. | For example, suitable for:<br>– Fine textured wall coverings<br>– Matt, non-textured paintwork / wall coverings<br>– Topcoat plaster, with grain size ≤ 1 mm as permitted by the respective manufacturer of the plasterboard system. | For example, suitable for:<br>– Smooth or textured wall coverings with gloss-effect, e.g. metal or vinyl wallpaper.<br>– Varnish or paintwork / wall coverings with semi-gloss effect.<br>– Stuccolustro or other high-quality smoothing techniques. | Suitability |
| Works include:<br>Stopping of butt joints between plasterboard panels and plastering over visible heads of screw fixings or similar.<br>Including application of joint tape (reinforcement strip) if included as part of the plastering system (joint filler, edge articulation of the panels). Insertion of joint tape also undertaken if required for constructional reasons. | Works include:<br>Basic filling (Q1) and subsequent filling / smoothing of joints to achieve a seamless transition between board surfaces.<br>No tool marks or trowelling ridges should be visible. If necessary, sanding of the trowelled surfaces.<br>The aim is to fill the joins so that there is a seamless transition between plasterboard panels. The same applies to fixings, internal and external corners as well as material junctions. | Works include:<br>Basic and standard filling / smoothing (Q2) including a wider fine trowelling of the joints and fine application of filler coat to remaining surface sufficient to close the material pores.<br>If necessary the trowelled surfaces should be sanded. | Works include:<br>Standard filling / smoothing (Q2) and wide fine trowelling of the joins as well as full surface application of smoothing coat including additional surface smoothing coat with an appropriate material (layer coat of up to 3 mm).<br>In contrast to Q3, the entire surface of the plasterboard wall is covered with a continuous smoothing coat / layer of plaster, for example by smoothing / filling the entire surface. Where necessary further preparatory surface treatment. | Application |
| | Traces of visible joints, especially when side-lit are possible. The effect can be lessened with a special filling coat (Q3). | Even after application of special filling coat, traces of ridges at joints still not entirely avoidable when side-lit but permissible according to VOB/C, DIN 18 350 (in future DIN 18 340), Nr. 3.1.2.<br>Amount and visibility much less than standard trowelling (Q2). | Possible traces from board panels and joints minimised, undesirable effects such as undulations and shadows or minimal marks largely eradicated.<br>The technical limitations of working on site should be noted. Trowelled surfaces that are absolutely flat and free of undulations when side-lit are not technically not feasible. | Limits |

## Trowelled gypsum plasterboard – dimensional tolerances

Table 4.5 : Dimensional tolerances for trowelled gypsum plasterboard surfaces

| Angular tolerances | Target sizes in m | | | | | |
|---|---|---|---|---|---|---|
| DIN 18202 1997-4 table 2, group 1 | up to 1 | from 1 to 3 | over 3 up to 6 | over 6 up to 15 | over 15 up to 30 | over 30 |
| | Permitted position deviations, in mm | | | | | |
| | 6 | 8 | 12 | 16 | 20 | 30 |
| Flatness tolerances | Distance between measuring points, in m, up to | | | | | |
| DIN 18202 1997-4 table 3 | 0.1 | 1 | 4 | | 10 | 15 |
| | Position deviations (limit values), in mm | | | | | |
| group 6: min. req. | 3 | 5 | 10 | | 20 | 25 |
| group 7: stringent | 2 | 3 | 8 | | 15 | 20 |

## 4.4 Earth plaster systems

### 4.4.1 Earth plaster systems in general

The selection of an appropriate plaster system depends on the substrate to be plastered and the desired surface quality. Other specific contributory factors include the degree of thermal fluctuation or the need to incorporate reinforcement fabric. The plaster system influences the choice of granularity and vice versa. The rules are similar to those of mortars with mineral binding agents, for example lime plaster mortar. As a rule topcoat plasters should not be stiffer than undercoat plasters, or else even slight deformations or thermal stresses in the undercoat will lead to the surface flaking.

Coarse-grain mortars (includes particles $\geq 1$ mm) are suitable for thicker coats of approx. 8 - 15 mm, in some cases as much as 35 mm. Fine-grain mortars (particles $\leq 1$ mm) are used for coats of up to 10 mm or for thin coats of 2 - 5 mm. The maximum permissible plaster thickness as well as the manufacturer's recommendations differ from product to product; always consult the respective manufacturer's instructions.

Single coat earth plasters require a sufficiently flat substrate with even absorbency properties. Surfaces that do not absorb evenly cause the plaster to dry unevenly which usually then becomes evident when the surface is worked. Grains that are already dry will rub off the surface more easily, for example over the underlying mortar joints.

A compromise between single and multi-coat plaster systems is to work over a thin preparatory levelling coat. These can be used to even out surface irregularities and lend the substrate more even suction characteristics.

High-quality surfaces are best realised using a multi-coat plaster system. Earth undercoat plasters usually serve as an excellent base for earth finishing plaster layers.

### 4.4.2 Shrinkage cracks in undercoat layers

Earth undercoat plasters are usually relatively rich mixtures. They should adhere well and provide a sufficiently stable base for subsequent plaster layers. As a result undercoat plasters have a greater tendency to form cracks than topcoat plasters. Smaller shrinkage cracks are not a problem for the adhesion of subsequent plaster layers. Large shrinkage cracks, however, can result in the undercoat plaster losing partial mechanical adhesion to the subsurface causing it to separate into patches that rise slightly at their edges (figure 4.4).

The degree of adhesion can be tested manually by applying thumb pressure to the edge of the crack. Loose particles should be removed and the cracks closed with a fill-

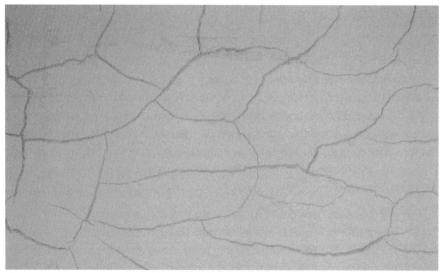

Fig. 4.4:
Critical degree of crack formation in an earth undercoat plaster

er of earth mortar before further plastering. Large shrinkage cracks otherwise show through to the surface coat.

Reasons for excessive shrinkage cracking include:
- Too rich clay mortar
- No fibrous reinforcement
- Too thick layers
- Mortars prepared with too much water
- Sandy, low suction or very smooth substrates (the substrate surface cannot help prevent shrinkage)
- The mortar has been worked too thoroughly and too long in a machine (causing more clay minerals to be activated in turn requiring more water)
- Too rapid drying and / or low suction of the substrate (causing shrinkage from the surface inwards)
- Very high suction substrates (causing shrinkage from the base outwards).

Before further plastering of the surface, the undercoat must be allowed to dry to the extent that no further crack formation takes place. Crack formation in the base coat after it has been plastered over results in cracking in the topcoat.

### 4.4.3 Reinforcement fabric

Unlike plaster laths, reinforcement fabric does not help the plaster adhere to the substrate. Its purpose is to resist tensile stresses within the surface of the plaster. Reinforcement fabrics help to reduce the degree of cracking to a tolerable level. While it is not possible to categorically avoid cracks forming, reinforcement fabrics minimise the risk.

The decision whether or not to use reinforcement fabric with earth plasters follows similar criteria to other plasters. Typical cases include:
- Irregular substrates that behave differently across their surface (for example: changes of material, hollow boxes for roller shutters, concrete lintels and ring beams)
- Soft substrates (for example: light earth, wood fibre insulation boards, reed mat insulation boards, straw bales)
- Surfaces that are subject to impact loads and vibrations (for example: the underside of timber ceiling joists).

Joint reinforcement tape is used to reinforce the joints between building boards, for example in the form of scrim tape for gypsum plasterboard panels.

It should be clear that reinforcement fabric is not able to resist more serious movement such as building settlement or deformations in the building's structural framework. In old buildings one should carefully consider whether it is worthwhile embedding reinforcement fabric in the plaster. In extreme cases, the presence of reinforcement fabric may mean that damages affect an entire surface whereas without reinforcement fabric smaller cracks might have formed.

As with other plaster systems, the reinforcement fabric should be placed 2/3 of the plaster thickness away from the plaster base, i.e. just beneath the surface. In the case of two-coat plasters, the reinforcement fabric is embedded in the surface of the undercoat layer.

The reinforcement fabric is laid directly onto the wet layer of plaster and carefully worked in with a felted or wooden float. The fabric must lie flat and taut and be free of creases. Laying reinforcement fabric onto dry substrates is not advisable as insufficient wet plaster is then able to penetrate it; in such cases the reinforcement fabric ends up being a separating layer that impairs the connection between the topcoat and undercoat plaster and may even cause the layers to separate (figure 4.5).

Individual strips of reinforcement fabric should be overlapped by at least 10 cm when applied to the whole surface. The overlap must be carefully worked to avoid the double layer acting as separating layer. The reinforcement mesh should extend into neighbouring areas beyond the area to be reinforced by another 25 cm. To reinforce shear

 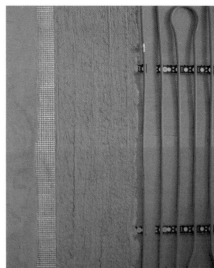

Fig. 4.5 (left):
Working in reinforcement fabric into the wet earth plaster surface

Fig. 4.6 (right):
The stages of plastering in pipes for wall heating

stresses that occur around openings for windows and doors, additional strips of reinforcement mesh can be placed diagonally across the corners of openings.

All kinds of reinforcement fabric with a mesh size large enough for the granularity of the mortar to pass through it are generally suitable. Reinforcement fabric with a mesh size of 5 mm is typically used for earth plasters. More coarse meshes usually have thicker fibres that are more rigid and cannot always be successfully worked into uneven or bulging surfaces. Hessian fabric, which is often used for earth plasters are more difficult to work than fabrics with dispersion-coated threads due to their fibrous threads. In concave mouldings and inside edges, hessian fabrics can separate from the layer beneath due to their tendency to shrink slightly. Casein-coated flax fabric meshes are well-suited for earth plasters.

### 4.4.4 Earth plaster on surfaces subject to thermal fluctuation

On surfaces subject to changing temperatures such as those with wall heating systems or stoves plastered with earth, fibre-reinforced plaster mortars along with the inclusion of reinforcement fabric are an effective combination. In the case of stoves, it may be necessary to clarify the fire protection requirements for the materials used before application (section 4.9.4).

The plastering of embedded heating pipes of wall heating systems is undertaken with the following steps (figure 4.6):

- Mount the heating register so that all pipes are enclosed by 10 mm of plaster. The pipework should be pressure-tested as per the rules of the trade prior to plastering.
- Plaster the heating pipes under operating pressure. Level off the plaster at the top-surface of the pipes.
- Dry the plaster as per protocol with heating on. The heating pipes are therefore at maximum expansion while the plaster is still viscous.
- Apply a second layer of plaster of 5-10 mm thickness over the pipes with the heating switched off. Directly after application, work the layer of reinforcement fabric into the still wet plaster. Overlap the edges of reinforcement fabric and extend it at least 25 cm into neighbouring surfaces that are not heated.
- After the plaster has dried fully, apply a final finishing plaster with the heating still turned off.

Ovens and stoves often have irregular forms and spherically curved surfaces. In such cases brick wire mesh is often used as a plaster base (section 4.3.4). Reinforcement fabric is always recommended for such cases but requires especially careful application.

## 4.5 Processing and application

### 4.5.1 Mortar preparation

If earth from a clay pit is to be used for the manufacture of site-made mortars, it must by crumbly and free-flowing in consistency. Numerous methods for achieving this have been documented in historical literature. Brickworks can generally provide earth in free-flowing form. When preparing site-made mortars, special care must be taken to ensure the components are well mixed.

Ready-mix mortar is generally mixed with typical mixing machinery. Small quantities can be mixed by hand or with a hand-held mixer, larger quantities with a tumble mixer or paddle mixer. Paddle mixers are more powerful than tumble mixer and the mixture is worked more thoroughly (section 3.2).

Fig. 4.7 and 4.8 :
PFT MULTIMIX compulsory mixer (batch mixer) and ZP3S mortar pump as separate pieces of equipment. Section through a similar model. See chapter 3.2 (source: PFT)

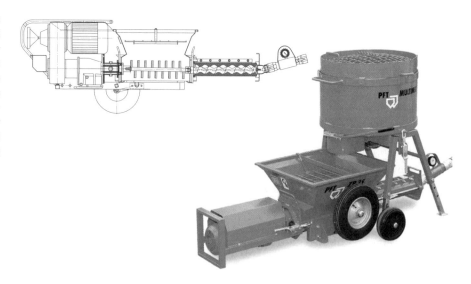

The workability of earth mortars is improved if the mortar is allowed to stand for a while after processing and then worked through once more. In the case of coloured earth plasters, this standing time is usually prescribed.

Naturally-moist mortars are processed with a *compulsory mixer* and *mortar pump*. This equipment is generally more powerful and the mixing method more effective at processing and achieving more stable mixtures but the cost of acquisition is relatively high and the machines are not very mobile on the building site. More mobile equip-

Fig. 4.9 :
PUTZMEISTER SP11 combined compulsory mixer (batch mixer) and mortar pump

Fig. 4.10 :
Worm drive mechanism (mortar pump) of the PUTZMEISTER SP11 (source: PUTZMEISTER)

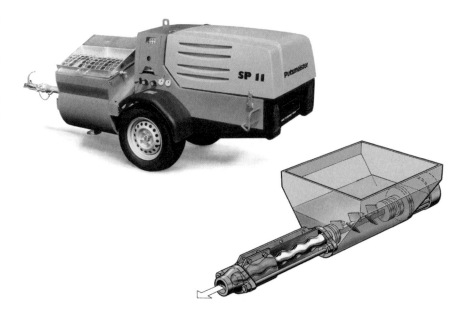

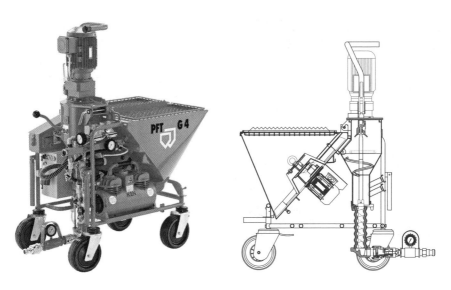

Fig. 4.11 : PFT G4 plastering machine (continuous mixer) with screw conveyer and mixing shaft (source: PFT)

ment has since come onto the market for processing naturally-moist mortars (figures 4.7, 4.8, 4.9 & 4.10).

Dry mortars can also be processed with *continuous mixers* such as normal gypsum plastering machines. The dry material is fed via a *rotary-vane* or *screw conveyor* into a *mixing chamber* with a *mixing helix*. The period in which the mortar remains in the mixing chamber and therefore the time in which it comes into contact with water is very short (figures 4.11 and 4.12).

Along with water contact, thorough mixing also contributes to better working properties and the stability of the earth plaster. For this reason the use of an *after-mixer* placed between the outlet of the plastering machine and the first mortar hose is recommended (figure 4.13).

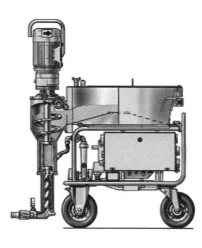

Fig. 4.12 : PUTZMEISTER MP25 plastering machine (continuous mixer) with star wheel drive and conical mixer (source: PUTZMEISTER)

Fig. 4.13:
After-mixer unit
beneath a PFT G4
(source: PFT)

Poor transport of the mortar in the mortar hose or poor workability as a result of only short contact with water should not be compensated for by adding extra mixing water. This makes the plaster thinner leading in turn to increased shrinkage cracking, poor stability and longer drying times.

The following table provides an overview of plastering machines and their application areas:

Table 4.6 : Suitable plastering machines and manufacturers for different kinds of earth plaster mortars

| | Manufacturer | Coarse plasters moist | Coarse plasters dry | Fine-finish plasters | Coloured plasters |
|---|---|---|---|---|---|
| 1 | PFT<br>www.pft.de | ● | ● | ● | ● |
| 2 | PUTZMEISTER<br>www.putzmeister.de | ● | ● | ● | ● |
| 3 | UELZENER<br>www.uelzener-ums.de | ● | ● | ● | ● |
| 4 | DEUTSCHE FÖRDERTECHNIK<br>www.deutsche-foerdertechnik.de | ● | ● | ● | |
| 5 | STROBL (by Putzmeister)<br>www.strobl-bc.de | | ● | ● | ● |
| 6 | M-TEC<br>www.m-tec-gmbh.de | | ● | ● | |
| 7 | INOTEC<br>www.inotec-gmbh.com | | ● | ● | ● |
| 8 | WAGNER<br>www.wagner-group.com | | | ● | ● |
| 9 | B & M<br>www.bm-vertriebs-gmbh.com | | | ● | ● |
| 10 | GRACO<br>www.graco.be | | | ● | ● |
| 11 | UEZ MISCHTECHNIK [1]<br>www.uez-mischer.de | ● | ● | ● | ● |

[1] Only mixing machines, no plaster pumps or spraying technology.

## 4.5.2 Mortar application

Earth plasters can be spread, thrown or sprayed. Thick mortars adhere better when thrown rather than spread. Historical literature recommends throwing as a more suitable means of application for coarse-grain earth plaster mortars [Niemeyer, 1946], and this method is also well-known for coarse-grain mortars with low adhesion. For earth plasters of normal consistency with sufficient adhesive properties, as well as for thin layer applications, plasters can be spread without any problems (figure 4.14).

In modern earth construction, however, the majority of earth plasters are applied by machine. The mortar is pumped from the machine in hoses by a *screw* or *piston pump* and propelled through a spray applicator with compressed air. Spray-applied mortar adheres better than hand-applied plaster.

Unlike other mortars, earth mortar can remain in the mortar hoses and machinery overnight or over the weekend because they are water soluble and do not cure. Clean surplus mortar does not need to be disposed of but can be reprocessed for later use (figure 4.15).

Although earth plasters are water soluble it is still necessary to protect other building elements and equipment against soiling. Porous surfaces in particular, such as light-coloured woods, are susceptible to staining if earth slurry is allowed to soak into their pores.

Fig. 4.14 (left) : Application of a thin layer of earth plaster mortar

Fig. 4.15 (right) : Spray application of earth plaster mortar

### 4.5.3 Corner profiles, plaster beads and junctions

The edges of earth plastered surfaces are usually rounded off. Alongside aesthetic considerations, a key reason is that rounded corners are less susceptible to denting or chipping. Special corner trowels and edgers for the easy finishing of outside and inside corners are available with or without corner radius or bevel (figures 4.16, 4.17).

Fig. 4.16 : Corner plaster profiles with large (radius preformed in the undercoat) and small corner radii, Japanese corner trowel with corner radius

Fig. 4.17 : Corner plaster profiles with large (preformed in the undercoat) and small corner bevels, Japanese corner trowel with corner bevel

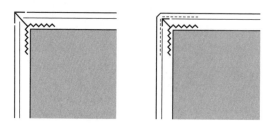

Fig. 4.18 : Corner bead with reinforced undercoat plaster and with corner protection for the topcoat plaster

Plaster beads and profiles for other kinds of plaster can also be used with earth plasters. One should, however, take into account that profiles made of corrosive materials will be exposed to moisture for longer and that when dry earth plasters will not protect them against further exposure to moisture.

If corner beads are used, they should be especially well fixed with fixing compound as earth plasters are comparatively soft and do not hold profiles in place as firmly as other plasters (figure 4.18).

The use of corner beads and plaster profiles with earth plasters can lead to crack formation due to the plaster's comparatively high measure of shrinkage. Plasters applied to profiles that are not removed after use should be reinforced with a strip of reinforcement scrim or joint tape (figure 4.19).

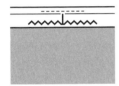

Fig. 4.19 : Undercoat profile with scrim reinforcement over profile

Casing beads, for example made of stainless steel, can be used to create clean edges of earth plaster surfaces. When using coloured earth plasters, care should be take to continue the undercoat plaster layer over the edge profile as the non-absorbent metal material will otherwise show through in the coloured topcoat plaster (figure 4.20).

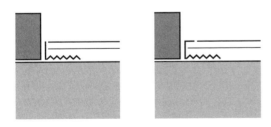

Fig. 4.20 : Casing beads (right: with flange) for creating a shadow line at the junction to another material

When plastering junctions between internal corners of rooms, it is important to take into account that while plastering the second surface, the first may be damaged at the junction between the two. Coloured earth plasters in particular must be masked at the corners, especially where the two walls have different colours. The wall must be fully dry before masking. Similarly, changes of colour on a flat surface are best achieved by carefully masking each work stage.

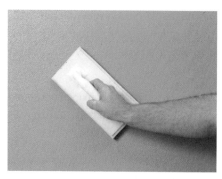

Fig. 4.21: Surface treatment with a sponged float and polishing trowel

### 4.5.4 Surface finishing

Working of the fresh mortar application, i.e. levelling, straightening with a featheredge or felting is undertaken in much the same way as with other plasters using typical plasterer's tools. Earth plaster surfaces are usually rubbed with a sponge, felted or wooden float. Smoothed surfaces are first finely rubbed and then smoothed using a polishing trowel or other special trowel (figure 4.21).

Many centuries of the traditional use of earth plasters in Japan have brought forth a series of special plastering tools that are particularly well suited for working the surface of earth plasters. Their quality lies in the way pressure is transferred from the handle to the blade. As a result very thin and flexible steel can be used with which it is possible to compress wet plasters to a very smooth and consistent finish (figure 4.22).

The air hardening characteristic of earth plasters requires a degree of experience and sensitivity in choosing the right time to work the plaster. This is much more dependent on the ambient climatic conditions and the suction characteristics of the substrate than with other chemically-curing plasters.

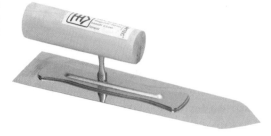

Fig. 4.22: Japanese earth plaster trowel

### 4.5.5 Special considerations for coloured earth plaster surfaces

Coloured earth plasters are not given any further treatment or coverings. For this reason the building site, all tools and all water used for mixing and washing must be clean. On the building site care must be taken not to damage finished surfaces: for example, all flooring work should already have been undertaken and workmen should observe the same level of care as they would for painted wall surfaces.

Coloured earth plaster surfaces should not be worked with too much water as the introduction of moisture into the thin layer of plaster may lead to shrinkage cracking in the same as it would if too much water were used in its mixing.

Homogenous coloured plaster surfaces can only be achieved if they have a uniform level of moisture content when worked. Different moisture levels influence the surface structure that results when working the surface, and the surface structure in turn dictates how the light reflects off smooth or rough surfaces. The importance of adequately preparing the substrate has already been discussed (section 4.3.5) but difficulties can also arise during the processing and application. For example, convection cycles caused by heating or draughts should be avoided as they lead to cracking as well as to differential patterns of drying. The application of coloured plaster to a continuous surface should undertaken swiftly without breaks until the entire surface is finished, likewise the subsequent working of the surface. Breaks in which a part of the surface starts to dry will be visible in the final finish.

As coloured earth plaster mortars consist of natural raw materials, slight colour inconsistencies within the material are possible. Sufficient earth plaster mortar must therefore be mixed to cover a continuous surface. The same applies for mixtures mixed on the building site with the manual addition of pigments. Slight colour variations are not as critical for adjoining surfaces that meet at a corner (e.g. corners of rooms, wall-ceiling junctions) as differences in light reflectance already make the surfaces look slightly different.

Even the best quality processing and application will not be able to prevent a slight degree of colour variation for all of the above reasons. In general this is tolerated – or even desired – by the client.

### 4.5.6 Drying

Due to the specific hardening characteristics of earth building materials, the drying process of earth plasters is dependent on the ambient climatic conditions and the suction characteristics of the substrate. To reduce delays in the construction and minimise prolonged levels of humidity during construction, it is necessary to ensure sufficient ventilation after plastering or, where this is not possible, to arrange machine-powered forced drying.

When scheduling building works, it is important to plan sufficient time for the drying of wet earth building materials. In the warmer months, these generally dry more quickly as warm air is able to hold more moisture than cold air. That said, cold winter air is more dry in absolute terms. If cold air is warmed, for example by moderate heating on the building site, it is able to absorb large quantities of moisture which can then be transported out of the building site. In summer by contrast, it is even possible that moisture can be transported indoors with fresh air from outside. When hot and humid air, for example from thunderstorms, passes over cool surfaces of the building, it can lead to the formation of condensation. At the same time, warm temperatures offer ideal conditions for mould formation.

In this respect, given the currently available possibilities for heating or cooling building sites, it is not winter that is the most problematic drying period but late summer and early autumn. During this period, and during winter when the building site is unheated, large quantities of air are required to ensure earth building materials dry adequately. Building sites that are heated in winter dry out comparatively quickly.

When not using machines, adequate ventilation must be achieved by natural means. Windows and doors must be left open day and night, ideally on opposite external walls to provide effective cross ventilation. An air change rate (the amount of times the entire volume of air is replaced per hour) of at least 4, sometimes much more, can be assumed for rooms with open windows. If doors and windows are closed the air change rate can sink to 0.8 or even less. Large quantities of air are required to dry out wet plaster surfaces. For example, a 2 cm thick layer of earth plaster applied to a surface area of 50 m² contains around 200 l of mixing water. A large proportion of this should ideally be able to dry within the first few days after application of the plaster.

Machine-powered forced drying is usually realised using *fan dryers* or *dehumidifiers*. Fan dryers increase the rate of natural ventilation. The fan should be placed with this in mind and air must be able to enter and leave the building. Without this the air simply circulates in the building but is not effective at drying. Simple and cost-effective machines can be hired and are able to shift between 100 m³ and 1,000 m³ or more per hour. Equipped with additional heaters, they can increase the moisture retention capacity of the air by an order of magnitude. The air should be able to flow unhindered past all wet building surfaces. A disadvantage with fans is that they also distribute a considerable amount of dust which can contain spores and nutrients for mould.

Condensation or refrigeration dryers function according to the principle of a heat pump. The water condenses on the cooling surface of a refrigerating compressor. Condensation dryers operate by recirculating the air; windows and doors must therefore be kept closed. The water removed from the air collects in water containers that must be periodically emptied. A condensation dryer can remove between 10 and 30 l of water in 24 hours. While this may not seem much given the amount of mixing water in the plaster mentioned above, in practice the plaster surface dries out relatively quickly. An advantage is that the drying process is gentle and uniform. At temperatures below 15 °C adsorption dryers should be used.

Machine-powered forced drying should be employed with restraint. Too rapid drying can result in stresses forming as a result of the top surface shrinking disproportionately to the lower-lying wetter layer. This risk increases proportionately the thicker the layer of plaster. In extreme cases, massive shrinkage cracks can cause a loss of adhesion in the plaster.

Additional moisture, for example from wet gypsum plasters and screeds can increase the drying load. In certain circumstances this can cause dry building materials, or building materials that have already dried, to acquire a critical level of moisture. This needs to be considered when scheduling building works. The drying of thick layers of plaster in modern timber-frame constructions can, for example, be a problem. The substrate generally exhibits low surface suction and the construction is generally well-sealed so that the air change rate is low. Wooden materials react to increased humidity by expanding or warping. For this reason, dry earth construction methods should be chosen for use in timber constructions (chapter 6).

If good ventilation cannot be guaranteed as a matter of course, the drying process needs to be organised by the contractor or another responsible person. The contractor should ideally offer to ensure responsible ventilation of the site, so that they can ensure it is put into effect when needed.

The *Lehmbau Regeln* recommend that the drying process is recorded using a drying protocol and the DVL Technical Information Sheet on Earth Plasters requires it for critical situations [DVL, 2008].

*Mould formation* on fresh earth plaster is almost always an indication of insufficiently fast drying. In unfavourable conditions (humidity, warmth), mould can form on all materials unless they have a very high pH-value or are treated with a fungicidal substance. The first generally whitish indications of mould are particularly easy to see against the brown background of earth surfaces, creating the impression that earth plasters are particularly susceptible to mould formation. This impression is misleading; many other plasters have a neutral pH-value but the white mould is barely visible against the white surfaces.

To clarify whether there really are clear indications that quantitative or qualitative aspects of mould formation are typical for earth building materials, tests were under-

Fig. 4.23: Example of a drying protocol sheet with explanations for use, Technical Information Sheet, DVL

## Drying protocol sheet

Project: ..................................................

Responsible supervisor ..................................................

Agreed drying measures

| Monitoring interval | Date/time | Drying measures undertaken | Drying progress | | | | | | |
|---|---|---|---|---|---|---|---|---|---|
| | | | | | | | | | |

### Purpose
A drying protocol is an instrument for formally monitoring and documenting the drying progress of earth plasters. It serves, among other things, to avoid or minimise the occurrence of temporary mould formation.

### Application
A drying protocol should be undertaken when
– layers in excess of 1.5 cm need to dry
– the plaster is applied to a substrate with poor absorbency (e.g. concrete)
– the construction site has a high relative humidity level (e.g. after laying wet screed).

### Drying measures
Drying measures include natural ventilation (cross-draught) or forced drying using machinery. The measures must be agreed among all participants in the building process and recorded in the protocol (e.g. "8 windows constantly open, 2 doors open for 10 hours per day" or "2 condensation dryers running constantly, windows and doors kept closed"). Drying measures should be chosen so that all plaster surfaces dry evenly. Caution: excessive use of forced drying (drying machinery) can cause stress cracking in the plaster!

### Responsible supervisor
The protocol should be undertaken by a responsible supervisor with sufficient expertise and experience. This can be a supervising architect, the earth building contractor, the client or another person in a responsible position.

### Monitoring protocol
The building site and the drying progress should be monitored at agreed regular intervals of not more than 48 hours. The monitoring serves to check that the agreed drying measures are in fact in place and to document the drying progress (for example "drying rapidly, first light patches appearing"). Deviations from the agreed drying measures should be communicated to the construction supervisor or planner immediately so that remedial measures can be implemented.

### Remuneration
Remuneration for monitoring and protocolling the drying process should be agreed.

taken from 2002-2004 at the *Institut eco Luftqualität + Raumklima GmbH* in Cologne [Röhlen, 2008]. They can be briefly summarised as follows:

▸ Tests of previously properly dried earth plaster surfaces after 28 days exposure under laboratory conditions to different levels of relative humidity of 40 - 60 % and > 95 % RH at a temperature of 23 °C (08/2002).

▸ Result: properly dried earth plaster surfaces subjected to normal conditions exhibit no indications whatsoever. Properly dried earth plaster surfaces subjected to very unfavourable conditions (23 °C, > 95 % RH) exhibit initial mould formation after 2 - 3 weeks, after 4 weeks significant mould formation. As such they behave similarly to other pH-neutral surfaces.

▸ Tests of plaster surfaces after 14 days exposure to poor drying conditions under laboratory conditions of 20 °C and > 95 % RH (01/2003).

▸ Result: increased bacterial content was found in earth plaster surfaces that were not allowed to dry properly. In products containing plant fibres, seeds germinated. The plaster was not covered with mould, only individual 'nests' of mould formed on the surface. In several cases of poor drying, not the plaster but the rear of the backing material exhibit mould formation. This was a cement-bonded wood-wool lightweight building board of the kind commonly used as a plaster base for many typical mortars.

▸ Tests of plaster surfaces made with mineral and fibre-reinforced mortars after 6 weeks exposure to poor drying conditions and very unfavourable hygienic conditions (09/2004).

▸ Result: of the earth plaster surfaces that were not allowed to dry properly over a period of 6 weeks and were exposed to very unfavourable hygienic conditions, the mineral plasters were on average less susceptible to mould formation than plant fibres. In some cases, these mortars also exhibited increased mould formation.

Conclusion: when properly allowed to dry, earth plasters do not exhibit a pronounced tendency for mould formation. The composition of earth plasters is, according to current findings, not a reason for the formation of mould but at most a contributory factor. Earth plasters without fibres additives must also be allowed to dry properly: spores are always present in the air and dust settling on fresh plaster can serve as a nutrient.

As soon as any mould formation is detected, the drying process must be accelerated and increased immediately. Time is a key factor in the formation of mould and the purpose of accelerated drying is to halt the growth of possible health hazards as quickly as possible. This should be followed through regardless of whether, once dry and covered over with subsequent layers, there is actually any risk of exposure for future residents or not.

An immediate measure for stopping mould growth is disinfection with 70 % alcohol or diluted hydrogen peroxide (work safety must be heeded!). As soon as the degree of drying is sufficiently advanced, the surface can be cleaned, for example with a vacuum-cleaner equipped with dust class H filter. The drying process should be more or less complete within a few days. Finally, the dry surface should be inspected for any remaining traces of mould structures, which if present should be removed. A coat of whitewash provides additional protection, particularly for the remainder of the construction period.

There is no evidence of mould formation on dry earth plaster surfaces. Even in critical situations such as cold bridges in existing historic buildings, earth is remarkably unsusceptible. This can be attributed to its open pore surface, the capillary conductive pore structure and, not least, its humidity regulating capacity.

### 4.5.7 Crack repairs and additional surface finishing

Shrinkage cracks in earth plaster surfaces can have several causes. Mortars may be insufficiently leaned down, the substrate may be uneven, the plaster mortar for a single layer application may have been prepared with too much mixing water or the plaster may have dried too quickly or too one-sided. In the case of coloured earth plasters, the mixed plaster may not have been allowed to stand sufficiently or too much water was applied when subsequently working the surface.

Shrinkage cracks can be minimised but they are not entirely avoidable without undue effort. In terms of fitness for purpose, fine cracks are permissible as long as they do not affect other technical requirements such as air tightness. Also permissible are hairline cracks (crack width ≤ 0.2 mm) in interior plaster surfaces of quality level Q2-rubbed [GIPS, 2003].

Smaller shrinkage cracks can usually be closed by wiping them closed with lightly wetted material from the surrounding surface. Larger cracks must be closed with additional material otherwise they will reappear once the plaster dries. Problems can arise when the plaster mortar is so coarse that it cannot be used to close the crack. In such cases the crack must either be sufficiently enlarged or a filled with a finer-grain mortar mass. The finer mass may, however, look different due to its different surface texture. Trials can be conducted on test surfaces to assess the potential quality of the repair. In the worst case entire surfaces may need to be given a thin skim coat application.

Cracks that have formed in an undercoat plaster that was not fully dry at the time of the application of the topcoat are more difficult to repair. Once they start showing through to the top surface there is little else one can do other than to apply a new coat of plaster to the entire surface, where necessary with embedded reinforcement fabric.

The repair of cracks in coloured earth plaster is particularly challenging (section 4.1.3). Trials should be conducted in advance on test surfaces to assess whether the potential results of the repair are likely to be satisfactory.

The brilliance of coloured earth plasters can be brought out by carefully wiping the surface with just a little water (wash out the sponge repeatedly while working). Alternatively the surface can be carefully polished with a soft wallpaper brush. This means of treatment also improves the degree of abrasion resistance and surface stability of the plaster. The plaster must, of course, be fully dry before working the surface.

Different kinds of colourless wax are also available for treating the surface of coloured earth plasters. The wax layer heightens the wear resistance of the surface and the colours appear more brilliant.

## 4.6 Paints and wall coverings for earth plasters

### 4.6.1 Painting and surface stabilisation

The composition, processing and application of earth plasters all aim to create a mortar structure that is as stable and durable as possible. For coloured earth plasters that receive no further treatment, these qualities are of central importance. Earth plasters that are destined to be painted should likewise have a sufficiently strong and abrasion resistant mortar structure. Provided the manufacturer does not stipulate another form of strengthening, mortars should correspond to mortar compressive strength group CS II according to the DIN EN 998-1, i.e. the compressive strength should be at least 1.5 N/mm² (section 4.7.1). Stronger plasters are more hard-wearing and less susceptible to damage. They are also easier to repair if necessary.

If earth plasters are to be painted or given a transparent fixative coating, the paint or fixative should not be film-forming and not have a too high binder concentration. Film-forming coatings have a tendency to peel and they hinder the sorption capacity and porous permeability of the surface. Coatings with a high binder concentration produce surface tension which can lead to cracking or peeling.

The following common paints and coatings are suitable for use with earth plasters:
- *Earth brush-on plasters* and *earth paints* are clay-based or lime-based paints (chapter 5). The paints have low tension characteristics and do not impede the breathability of the surface.
- Pure *limewash* should be heavily diluted and applied in many layers. The earth plaster should be gently pre-wetted. Even with careful execution, surface chalking cannot be entirely ruled out.

- *Lime-casein paints* must be properly mixed or sufficiently diluted before use. They must be carefully processed to eliminate the risk of tension cracking. The vapour permeability of the plaster is slightly impaired by the paint.
- *Mineral paints* can only be used with earth plasters in the form of *dispersion silicate paints* and some paint manufacturers advise against their use. Dispersion silicate paints provide a hardwearing surface. Because they contain acrylate their use as a coating for natural building materials is sometimes viewed critically. Dispersion silicate paints impair the vapour permeability of the plaster slightly more than other paints do.
- *Glue-bound distemper, chalk-casein paints* and *resin-based dispersion paints* are also suitable for use with earth plasters. Other paints are also possible.

Depending on the paint system used or the manufacturer, it may be necessary to first prime the plaster surface. Usually it is sufficient to apply an undercoat of diluted paint.

The number of coats necessary depends on the paint used, and of course on the desired quality of the end result. The lighter the colour of the paint, the more coats will be necessary, especially on dark plaster.

Best results are achieved when applying paints with a wide brush rather than a roller. While the roller is time-saving, the surface texture it produces obscures that of the earth plaster to an unacceptable degree (section 5.4).

If no paint is desired, it is possible to fix the surface with a heavily diluted transparent potassium water-glass solution. It is recommended to apply multiple coats as too much binder in a single coat can create surface tension in the stabilised surface. It is advisable to conduct trials on test surfaces in advance. If prescribed by the manufacturer, fixatives should be used as instructed.

### 4.6.2 Wallpaper

Earth plasters are generally not wallpapered so as not to impair their advantageous vapour-permeable surface characteristics. If surfaces need repeated application of new wallpaper (e.g. for rented apartments), the water solubility of earth plasters can present a problem.

It is nevertheless possible to wallpaper earth plasters. If the plaster is well prepared with a well-conceived plaster build-up, the wallpaper can also be removed again. The earth plaster must first be stabilised with a primer, then smoothed with a skim coat of suitable filler.

### 4.6.3 Lime skim coat plaster

Historically, earth plasters were commonly coated with lime-wash, fine-sandy lime slurry (lime whiting) or thin layers of lime plaster (section 10.6.1). Today thin skim coats of lime plaster are usually applied as part of conservation measures. As a rule, light-coloured earth plasters are a better match for earth undercoat plasters. Thin lime plasters do not adhere so well and are more susceptible to flaking. If a lime skim coat is to be applied, the surface should first be prepared with a sufficiently adhesive lime slurry carefully worked into the pre-wetted earth. A finer ca. 3-4 mm thick skim coat of lime plaster is then applied. Lime mortars that are soft, such as air-hardening lime or slaked lime are most suitable.

### 4.6.4 Wall tiles on earth plasters

Wall tiles require a stable base. In particular where surfaces are regularly exposed to water at different temperatures, a sufficiently stable base for tiling such as a lime or cement plaster or plasterboard for use in wet rooms are necessary. Such areas include surfaces in showers or around baths.

Where wall tiling is desired in other less demanding situations, earth plasters can serve as a base provided the manufacturer has not explicitly ruled out their use in domestic kitchens and bathrooms or as a base for simple tiling applications. In most cases, the earth plaster must be stabilised, for example with a deep-penetrating primer. The primer solution must be able to penetrate deep below the surface (test in advance!). The use of primers that only stabilise the surface is problematic as it can lead to large differences in stress between the upper, stabilised layer and the lower, non-stabilised layer.

In general, bathrooms or kitchens in which earth plaster has been used are tiled only in localised areas as the advantage of the earth plaster is its ability to buffer and equalize the moisture in the room.

## 4.7 Requirements of earth plasters

### 4.7.1 Mechanical properties

The basic required mechanical properties of earth plaster mortars are given in the *Lehmbau Regeln* [DVL, 2009]. These should be defined and declared during the planning process, in particular with regard to the desired strength.

Plasters should adhere evenly to the substrate and the individual layers to one another. Within each layer, the mortar should have a consistent structural stability.

Earth plasters for typical requirements such as for coatings or wallpapers should be made with earth mortars of compressive strength mortar group CS II (compressive strength 1.5 - 5.0 N/mm²) according to DIN EN 998-1.

The measure of linear shrinkage of the plaster mortar should generally not be more than 2 %. Plaster mortars for thin skim coat applications can exhibit a higher measure of shrinkage. The measure of shrinkage can deviate from this provided it does not compromise its fitness for use and the appearance of the plaster. Fibre additives can contribute to limiting the degree of crack formation.

The table shows the requirements that all earth mortars are expected to fulfil. In the case of *factory-made mortars* and *semi-finished factory-made mortars*, the fulfilment of the requirements must be documented and the values declared. In the case of *site-made mortars* equivalent requirements apply in as far as they are required for the intended use. The craftsman in charge is responsible for their fulfilment. Instead of ascertaining the measure of shrinkage, the shrinkage (susceptibility to cracking) can be evaluated more simply using an in-situ test.

Supplementary required mechanical properties of earth plaster mortars, earth plaster systems and surfaces are defined in the DVL Technical Information Sheet TM01 *Anforderungen an Lehmputze (Requirements of earth plasters)* [DVL, 2008].

This stipulates that earth plaster systems and surfaces should be able to sufficiently and durably withstand typical wear and tear and be able to maintain their physical properties in the long term.

With regard to the requirement given in the *Lehmbau Regeln* concerning the adhesion of the layers to the substrate, the Technical Information Sheet TM01 notes that small-scale hollow areas do not represent a diminution of the fitness for purpose provided that the overall stability of the entire plaster structure is satisfactory.

Table 4.7 : Required mechanical properties of earth plaster mortars, Lehmbau Regeln [DVL, 2009]

| Bulk density (hard mortar) [1] | Compressive strength [2] N/mm² | Measure of shrinkage [3] % |
|---|---|---|
| defined | > 1.5 | ≤ 2 |

[1] The bulk density must be specified in kg/m³ rounded off to the nearest 100 kg. According to Table 5.1 in the Lehmbau Regeln, the dry bulk density of earth mortars lies between 600 and 1800 kg/m³.
[2] Mortars for plasters intended for use as a base for coatings and wallpapers.
[3] With fibrous additives, a thin coating and careful subsequent surface treatment as well as for small surfaces, a measure of shrinkage of more than 2 % may not present any problems.

In addition, the Technical Information Sheet TM01 stipulates that in general use the layer integrity must also be guaranteed, where applicable, to withstand reasonable vibrations and distortions of the substructure.

Earth plasters must, in accordance with the proposed use, possess sufficient firmness and resistance to abrasion.

For *factory-made mortars* and *semi-finished factory-made mortars*, manufacturers are recommended to undertake abrasion resistance and adhesion strength tests and to declare these in their product literature. The abrasion resistance of earth surfaces is determined according to the process developed by the *Forschungslabor für experimentelles Bauen* (FEB Building Research Institute) at the University of Kassel in Germany. The adhesion strength can be tested according to DIN EN 1015-12:2000 (section 4.8.1).

Table 1 of the TM01 concerns the plaster mortar, i.e. the building material. The required compressive strength of earth mortars is specified according to the intended use. For secondary spaces, the TM01 reduces the requirements given in the *Lehmbau Regeln*. Similarly, the required compressive strength of mortars for plasters whose surfaces are stabilised is reduced. Alongside earth plasters destined to serve as a base for coatings, the table also covers earth plasters whose surfaces are left untreated and exposed, for example coloured earth plasters.

Table 2 of the TM01 concerns plaster systems and surfaces, i.e. the building element. Here too the requirements are specificied according to intended use, this time for adhesive strength and abrasion resistance. The table differentiates between *technical test procedures* and *hand assessments*. Test procedures need only be undertaken

Table 4.8 : Minimum requirements for the compressive strength of earth plaster mortars in relation to their intended use. Table 1 of the DVL Technical Information Sheet TM01 [DVL, 2009]

| Intended use | Compressive strength N/mm² |
|---|---|
| Secondary spaces | ≥ 0.5 |
| Earth plaster for subsequent surface stabilisation in rooms for general use, e.g. living and working rooms in houses and apartments | ≥ 1.0 |
| Earth plaster as a base for topcoat plaster, coatings and wallpapers | ≥ 1.5 |
| Earth plaster to be left as a fair-faced non-stabilised surface in rooms for general use, e.g. living and working rooms in houses and apartments | ≥ 1.5 |
| Earth plaster to be left as a fair-faced non-stabilised surface in rooms subject to higher wear and abrasion (e.g. public facilities) | ≥ 1.5 [1] |

[1] *The suitability of the material for the respective purpose must also be individually assessed by conducting trials on sample surfaces before commencing application.*

where there is reasonable doubt regarding the material's fitness for purpose. A hand assessment should be undertaken by the contractor on site, in particular in the case of site-made mortars as these are not subject to the same declarative obligations as factory-made building products.

## 4.7.2 Building biology requirements

Earth plasters should not contain hazardous contaminants. At the time of writing, a statutory list of natural constituents, or for that matter additives, that may present a possible health hazard along with tolerable levels of concentration does not exist. In the absence of official guidelines, the *Natureplus Issuance Guidelines 0803 Earth Plaster Mortars* [Natureplus, 2005] provide some degree of orientation.

Table 4.9 : Requirements for earth plaster applications and surfaces on the building element. Table 2 of the DVL Technical Information Sheet TM01 [DVL, 2009]

| Intended use | Technical test procedure | | Hand assessment | |
|---|---|---|---|---|
| | Adhesive strength according to DIN EN 1015-12:2000 established on site $N/mm^2$ | Abrasion resistance according to Minke, established on site $g$ | Adhesion, pressure and bending tensile strength, established on site Finger tip pressure | Abrasion resistance, tested on site Wiping with the hand using light pressure |
| Earth plaster surfaces in secondary spaces | – | – | With repeated light pressure, no flaking or collapse of clods | – |
| Earth plaster, also after surface stabilisation, as a base for topcoat plasters, coatings and wallpapers | ≥ 0.03 [1] | ≤ 3 | With repeated medium pressure, no flaking or collapse of clods | Rubbing off and medium sanding acceptable [2] |
| Earth plaster, also after surface stabilisation, to be left as a fair-faced non-stabilised surface in rooms for general use, e.g. living and working rooms in houses and apartments | ≥ 0.03 [1] | ≤ 1 | With repeated medium pressure, no flaking or collapse of clods | Slight rubbing off acceptable. Only individual sand particles may loosen [2] |
| Earth plaster, also after surface stabilisation, to be left as a fair-faced non-stabilised surface in rooms subject to higher wear and abrasion (e.g. public facilities) | ≥ 0.03 [1] | ≤ 0.5 | With repeated medium pressure, no flaking or collapse of clods | Slight rubbing off acceptable. Sanding virtually ruled out [2] |

[1] To the substrate as well as between and within the individual layers of plaster.
[2] Where surfaces have been rubbed or felted, the assessment should be undertaken after brushing down the surface.

The guidelines state first of all that earth plasters that are declared as non-stabilised must consist of 100 % mineral and renewable raw materials. The only permitted binding agent is clay or earth.

More specifically, earth plaster mortars may not contain *biocide contaminants, halo-organic compounds* or *synthetic materials and fibres* such as acrylate or polyvinyl acetate. The total volatile organic compounds (TVOC) present in dry earth plaster mortar is limited to a maximum of 100 ppm. A pH value of < 8 is specified.

Table 4.10 : Maximum permissible values for the natural constituents of earth plaster mortars according to Natureplus Issuance Guidelines 0803 – Earth plaster mortars

| Substance or effect | Maximum value  mg/kg |
|---|---|
| Arsenic (As) | ≤ 5 |
| Cadmium (Cd) | ≤ 1 |
| Cobalt (Co) | ≤ 20 |
| Chromium (Cr) | ≤ 20 |
| Copper (Cu) | ≤ 35 |
| Mercury (Hg) | ≤ 0.5 |
| Nickel (Ni) | ≤ 20 |
| Lead (Pb) | ≤ 15 |
| Antimony (Sb) | ≤ 5 |
| Tin (Sn) | ≤ 5 |
| Zinc (Zn) | ≤ 150 |
| Organic contaminant proportion TVOC | ≤ 100 [2] |
| VOC classified in: K1, K2; M1, M2; R1, R2 MAK III.1 and MAK III.2 [3] | not detectable [1] |
| AOX | ≤ 1 |
| pH value | ≤ 8 |
| Artificial radioactivity: Cs-137 [4] | not detectable [1] |
| Natural radioactivity: [4] cumulative value according to ÖNORM S 5200 | ≤ 0.75 |

[1] *Determination limit: 1 mg/kg.*
[2] *Content in dry earth plaster mortar.*
[3] *K = carcinogenic; M = mutagenic; R = toxic for reproduction; Classification according to GefStoffV (D), Ordinance on Hazardous Substances.*
[4] *Detection of the activity in Bq/kg of the radioactive nuclides K-40 and Cs-137 as well as the Th-series, U-series and Ac-series using a gamma spectroscopy detection limit of 0.5 Bq/kg.*

Maximum tolerable values are also given for theoretically possible constituents with possible toxic effects, as well as for natural radioactivity. Even when such concentrations practically never occur, from the point of view of consumer protection it is nevertheless welcome when such exclusions are scientifically tested and documented.

### 4.7.3 Moisture sorption capacity requirement

Earth plasters should adsorb and desorb airborne moisture significantly more quickly and to a greater degree than other plasters (section 3.3).

### 4.7.4 Visual requirements

The visual requirements of a plaster are defined by the desired surface finish and the planned further treatment of the surface. The visual requirements can be described with reference to the *Quality Classification of Internal Plaster Surfaces* as given in DIN V 18550 2005-4 taking note of the aforementioned reservations (section 4.3.5).

The best way to describe the desired quality of surface finish is to plaster a sufficiently large sample area directly on site and under documented, realistic lighting conditions. This sample should fulfil the respective individual requirements with regard to colour, texture and smoothness and be agreed with the client. The sample area can then be used as a reference surface for all other subsequent plastered surfaces. This is particularly important for surfaces that are not destined to be treated with further coatings that affect its visual appearance.

## 4.8 Guaranteeing material properties

### 4.8.1 Basic testing procedures and declarations

Basic testing procedures and the properties that need to be declared for earth plaster mortars are specified in the *Lehmbau Regeln*.

In general, the constituents of factory-made products should be openly declared.

Furthermore, for *factory-made mortars* and *semi-finished factory-made mortars*, the properties must be clearly and legibly marked on the container or on an information sheet provided with the container.

### Bulk density

*Test:* test cubes with an edge length of 10 cm are made using the same procedure as used on the building site. Alternatively the cuboid blocks used for testing compressive strength (see below) can be used. The rounded mean value of at least three tests should be taken. Individual values may not deviate by more than 10 % from the mean value.

*Declaration:* Value in kg/m³ rounded to the nearest 100 kg.
Permissible deviations: > 600 and ≤ 1200 kg/m³ +10 %, > 1200 and ≤ 1800 kg/m³ +5 %.

### Compressive strength

*Test:* The compressive strength is measured in accordance with DIN EN 1015-11 on three test prisms with the smallest value taken as the end result. The test specimens are to be made with mortar of a consistency as made for normal working on site.

*Declaration:* Minimum measured value in N/mm².

### Measure of shrinkage

*Test:* The linear measure of shrinkage is tested on three test specimens measuring 160 × 40 × 40 mm. After removal from the mould, the test specimens are laid on foil and allowed to air dry. The final measurement is then taken. The test specimens are made with mortar with a consistency of 175 mm slump diameter, in accordance with DIN EN 1015-3.

*Declaration:* Maximum measured value in %.

## 4.8.2 Supplementary testing procedures and declarations

Further supplementary tests and declarations for earth plaster mortars are proposed in the DVL Technical Information Sheet TM01 *Anforderungen an Lehmputze* [DVL, 2008].

### Adhesive strength

*Test:* Test procedure as described in DIN EN 1015-12:2000.

*Declaration:* Minimum measured value in N/mm².

### Abrasion resistance and necessary surface hardening

*Test:* The abrasion resistance of earth surfaces is measured using a procedure developed by the *Forschungslabor für Experimentelles Bauen (FEB Building Research Institute)* at the University of Kassel in Germany. For this purpose a hard rotating brush of 7 cm diameter is pressed against the earth surface with a pressure of 2 kg and the resulting wear debris after 20 rotations recorded in grams. For felted or rubbed surfaces the test should be conducted after the surface has been brushed down.

*Declaration:* The degree of abrasion of the rubbed or felted surface should be declared. Manufacturers of earth plasters with an abrasion of more than 3 g should declare in the product information that to achieve a surface fit for purpose, surface stabilisation is necessary.

### Stabilisation

Stabilised earth plaster mortars must be declared as such. The kind and proportion by mass of the stabiliser additives must be given.

### Humidity sorption

*Test:* The sorption of earth plasters is measured using at least 250 cm² large and 15 mm thick earth test specimens that are sealed on five sides so that sorption can only occur on one face. The test specimen should be stored in a climatic test chamber at 23 °C and 50 % RH until it reaches its equilibrium moisture content. The relative humidity should then be raised as quickly as possible to 80 % and the increase in weight of the test specimen recorded at predefined intervals. The measurements are to be taken with weighing scales accurate to 0.1 g on at least 3 test specimens. The mean value of the 3 measurements is taken as the test value. If an individual measurement deviates by more than 20 %, it should not be included. For this reason it is generally recommended to prepare 5 test specimens.

The testing of thin layer applications of plaster requires the application of the plaster to an undercoat plaster to achieve the necessary 15 mm thickness of the test specimen. The undercoat plaster used should be stated.

Where certain products recommend surface stabilisation of the plaster, the humidity sorption should be ascertained using samples that have been treated with the recommended fixative or stabilising solution.

### 4.8.3 Quality control

According to the *Lehmbau Regeln*, *factory-made mortars* and *semi-finished factory-made mortars* must fulfil the general requirements as well as the declared values. For *site-made mortars*, responsibility lies with the respective craftsman to ensure the required properties of the plaster.

According to DIN EN 998-1, third party inspection is not necessary for plaster mortars and quality control is a matter for the manufacturer.

Declarations regarding the building material class (fire performance) of mortars are, by contrast, subject to monitoring by an approved body.

## 4.9 Building material and building element properties

### 4.9.1 Mechanical properties

Table 9.11 : The spectrum of mechanical properties of earth plasters

|  | Compressive strength N/mm² | Adhesive strength N/mm² | Abrasion g |
|---|---|---|---|
| from | 0.7 | 0.03 | 6.7 |
| mean | 1.5 | 0.05 | 3.0 |
| to | 2.5 | 0.15 | 0.3 |

### 4.9.2 Thermal insulation and protection against moisture

Commonly used earth plasters have a bulk density of between 1400 and 1800 kg/m³. To achieve the same level of insulation as other typical thermal insulation plasters, $\lambda \leq 0.1$ W m/K, a bulk density of $\leq 300$ kg/m³ would be needed, which can only be achieved by adding a considerable amount of lightweight aggregates to the mixture. This, however, is not possible with currently available technology. As such, commercially available "thermal insulation plasters" made with earth plaster should be regarded with caution.

Table 9.12 : Bulk density, λ-value and μ-value of earth plasters

| Bulk density kg/m³ | λ-value W m/K | μ-value |
|---|---|---|
| 1000 | 0.35 | 5/10 |
| 1200 | 0.47 | 5/10 |
| 1400 | 0.59 | 5/10 |
| 1600 | 0.73 | 5/10 |
| 1800 | 0.91 | 5/10 |

Air tightness

With regard to the air tightness of plasters, DIN 4108-3 *Protection against moisture subject to climate conditions; Requirements and directions for design and construction* in chapter 6 states the following: "Facing brickwork and timber-frame structures as well as brick masonry (in accordance with DIN 1053-1) are on their own not airtight in the sense of this requirement; such wall surfaces must be faced with a layer of plaster (in accordance with DIN 18550-2) or given another appropriate airtight treatment. Airtight in the sense of this requirement are, for example, concrete elements manufactured in accordance with DIN 1045-1 and DIN 1045-4 or plasters in accordance with DIN 18550-2 or DIN 18558."

DIN 4108-7 *Air tightness of buildings – Requirements, recommendations and examples for planning and performance* in section 5.2.1 is more general in its specification: "In order to achieve a sufficient level of air tightness for masonry constructions it is generally necessary to apply a layer of plaster."

Earth plasters are not mentioned explicitly in DIN 18550-2. With regard to their effectiveness as an airtight layer, the proposal made here takes a practical point of view: earth plasters with a bulk density ≥ 1400 kg/m³ exhibit a dense structure similar to other plasters. As DIN 18550-2 does not prescribe any plaster thicknesses, earth plasters can be seen as an airtight layer when they are continuous and do not exhibit any cracks, with the exception of very fine hairline cracks of ≤ 0.2 mm thick.

## 4.9.3 Sound insulation and acoustics

In the context of sound insulation, plasters serve as a continuous layer to prevent the occurrence of leakages, for example in masonry wall constructions. Apart from this function, when applied in typical thicknesses they have little influence on the sound insulation of building elements. For the classification of building elements according to DIN 4109, values for mineral plasters of a comparable bulk density should be used for orientation.

The acoustic properties of plasters depends on their hardness and the pore structure of the surface in particular, which can be influenced, for example, by the presence of fibres in the mixture. At the time of writing there are no acoustically-optimised earth plaster products on the market. Likewise, measurements of the reverberation time or other acoustic properties are at present not available for standard earth plaster products.

### 4.9.4 Fire performance

Building material class

Earth blocks and mortar *without* fibrous additives are classified in the currently valid DIN 4102 Part 4 (March 1994) as the equivalent of *non-flammable*.

When earth mixtures contain organic additives, they are not classified (in accordance with DIN 4102-4). According to DIN EN 998-1 mortars are, however, generally regarded as *non-flammable* when "…their proportion of homogeneously distributed organic material constitutes no more than 1% of the mass or volume (whichever is larger)". Plasters with a greater proportion of organic material should be classified according to DIN EN 13501-1. Details regarding the flammability of straw-clay mixtures are given in the *Lehmbau Regeln*. Tests conducted according to standard procedures have demonstrated that mixtures of earth and straw with a bulk density > 1200 kg/m³ exhibit the equivalent material class properties of *non-flammable* building materials.

Fire resistance class

DIN 4102-4 details the fire resistance classification of numerous building elements coated with plaster. The definition of "plaster" given therein refers repeatedly to DIN 18550. However, when the currently applicable edition of DIN 4102-4 was introduced, earth plasters were not yet detailed in DIN 18550. For the purposes of applicability by analogy, the crystallised water content is relevant for determining the fire resistance. Earth contains a comparable amount of crystallised water content to gypsum. Moreover, when exposed to fire, its structure does not disintegrate but actually becomes harder by forming ceramic structures. The mechanical adhesion of the plaster to the underlying surface can be ensured, for example, through the use of a plaster lath.

# 5 Paints and finishes

## 5.1 Terminology, composition and applications

Earth surface coatings such as paints and brush-on plasters are essentially clay-bonded materials. Commercially available products, however, also include a further binding agent for stabilisation, such as starch or cellulose. The colour of the finish is a product of the clay or earth use, or results from the addition of pigments.

As paints are very thin coatings, only small amounts of clay minerals are spread onto the painted surface. As a consequence, earth surface coatings exhibit negligibly small water vapour sorption characteristics (section 3.3). They are, however, vapour permeable, allowing moisture through to the underlying layer.

Earth surface coatings are water soluble and in terms of their use are comparable to distemper. It is important to take into account the water solubility when applying renovation coats (see section 5.5 below). Earth surface coatings are less hard-wearing than dispersions or other paints that harden. They do, however, have a very low surface tension and are therefore well-suited for soft substrates such as earth plaster.

Brush-on plasters are paints with a granular consistency. They can be used to create surface finishes that resemble finely rubbed plaster.

## 5.2 Substrate preparation

The substrate should generally be *stable, dry, free of dust* and *sufficiently rough* to provide a good key. Any remaining wallpaper or loose paint should be removed. Areas with *substances that may show through* should be sealed or primed using suitable painter's products. Special care should be taken with tar or nicotine stains.

The *wall suction* (rate of absorption) and *surface flatness* of the substrate should also be as even as possible. This equates to the surface quality level classifications Q3 or Q4 as described in the classification system published by the *Bundesverband der Gipsin-*

*dustrie e. V. (German Association of the Gypsum Industry)* for the preparation of surfaces ready for painting with earth paintwork or finishes (see section 4.3.6).

## 5.3 Priming

In many cases surfaces are primed to reduce and even out the wall suction characteristics. Priming is recommended for all substrates made of earth materials to avoid the earth paint or finish from being discoloured or soiled by earth washing out of the underlying surface. Instead of using plastic-based primers, one can, for example, use casein-based primers which offer a limited degree of waterproofing. A primer of this kind also simplifies the application of later renovation coats on earth substrates.

Light-coloured paints applied to dark surfaces do not appear as light. It is, therefore, advisable to lighten dark surfaces with a light undercoat, for example by using a white primer.

If a granular brush-on plaster is to be applied on top of a granular primer coat, the texture of the primer coat should match the desired texture of the subsequent top coat of brush-on plaster, for example through crosswise application.

## 5.4 Mixing and application

Earth paints and finishes must be allowed to swell for a while after initial stirring and then worked through once again before application. They are usually applied with a brush or wide painter's brush, often following the swing of the brush in a horizontal figure of eight shape or crosswise (see figure 5.1).

While it is quicker to apply paints with a roller, the resulting surface finish may not meet the client's expectations. It is worth taking the time to clarify the desired means of application in advance. When working brush-on plasters, one should take into account the fact that the granular constituents sink to the bottom of the paint pot more quickly than they do for example with dispersion paints. The paint should therefore be mixed more often, taking care to mix in settled mass from the bottom. During the application of the paint, excessive heating and draughts should be avoided.

The number of coats to be applied depends on the type of substrate and the desired end quality of surface finish. Typically two coats are needed on top of the pre-applied primer coat.

Fig. 5.1 : The crosswise application of earth brush-on plaster

## 5.5 Renovation coats

When applying renovation coats, it is important to take into account the water solubility of the underlying earth finish. Depending on the intensity of the new coat of paint, the surface below may be wettened to a greater or less degree, causing it to mix with the new coat of paint. This is particularly important to consider when the renovation coat has a different colour to the old earth surface; in such cases it may be advisable to first apply a fixative. Water soluble earth paints and finishes can, if required, be washed down from the surface. This is a further reason why earth substrates should be primed as described above.

# 6 Dry earth construction

## 6.1 Introduction

The use of dry earth construction techniques and building materials has increased significantly over recent years. In the past, the regional availability of suitable earth with good working properties were the principal criteria for using earth. Today, earth construction has to address quite different demands. In modern construction, tight time schedules and less wet construction are important criteria. In response to these requirements, earth building techniques have been developed that can be installed by carpenters and drywall contractors. In particular, the development of prefabricated panels and building boards has made it possible to use earth materials for the entire spectrum of dry construction techniques. Prefabrication also makes it possible to develop building materials that, unlike wet earth building materials, are not subject to drying shrinkage. A high proportion of moisture-absorbent clay minerals in dry building materials is not a problem; on the contrary, they can be used to create areas with especially intensive "moisture absorbers".

## 6.2 Clay panels

Panel-shaped building materials made of earth are used for internal partitioning walls, wall linings, suspended ceilings and the lining of roof spaces. They can be used as boarding over a grid-like subconstruction or as dry plaster panels applied to flat surfaces such as concrete or lime-sand blockwork, existing plastered surfaces as well as massive timber constructions or wood particle boarding. Earth panels are also available as *wall heating elements* with embedded heating loops or as *hypocaust elements* with integral ducts for warm air circulation. Thicker panel elements are also available that can be bonded or laid like bricks; their application is described in chapter 8: *Earth block masonry*.

Clay panels are made of clay or earth mixtures in combination with mineral and organic additives. They can have a homogenous structure or contain embedded stabilising reinforcement such as reed matting or loops of heating piping.

  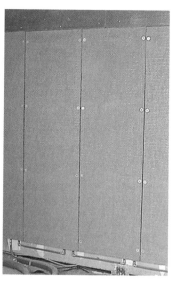

Fig. 6.1 : Clay panels as plasterboarding — Fabric coated panels are made industrially, either on an assembly line or individually. Panels made using an extrusion process can be given a perforated cross-section (longitudinal channels).

Fig. 6.2 (centre) : Clay panels as dry plaster tiles

Fig. 6.3 (right) : Clay panels as wall heating elements

The following provides a brief overview of a selection of widely available products on the market (as of 2008):

CASA NATURA clay building panel

Composition: earth, sand, sawdust. Bulk density: 1440 kg/m³. Production: extrusion. Format: 100 × 25 × 2.5 cm. Subconstruction spacing: 33 cm.

CLAYTEC clay plasterboard

Composition: earth or clay, organic and mineral additives. Bulk density ca. 700 kg/m³. Reed matting reinforcement, hessian facing. Production: conveyor assembly line. Lightweight building board for wall lining, partitioning walls, suspended ceilings, roof space lining. Formats: 150 × 62.5 × 2.0 or 2.5 cm, 62.5 × 62.5 cm × 1.6 mm as dry plaster tiling panel. Subconstruction spacing: 37.5 cm for 2.0 cm thick boards and 50 cm for 2.5 cm thick boards. Subconstruction spacing for ceilings and roof inclines: 37.5 cm for 2.0 and 2.5 cm thick boards.

HOCK clay building panel

Composition: earth, sawdust. Bulk density: 1600 kg/m³. Production: extrusion. Format: 100 × 25 × 2.5. Subconstruction spacing: 33 cm.

## 6.2 CLAY PANELS

### HYPOTHERMAL earth wall heating element

Earth wall heating element for wall linings with twin vertical ducts for warm air circulation *(hypocaust warm air heating)*. Production: extrusion. Surface area: approx. 2/3 of the floor area of the room to be heated at a height of max. 2 m and 40 - 50 °C supply temperature. Panel format: 100 × 29 × 10 cm; U-shaped elements to connect the supply and return flow at the top and bottom of the panels: 7 × 58 × 7 cm. The elements are typically bonded together with adhesive mortar.

### LEBAST clay boards

Composition: earth, chopped straw. Bulk density: 1200 kg/m³. Glass-fibre fabric facing. Production: individual pressed. Formats: 125 × 62.5 × 2.2 or 1.4 cm. Subconstruction spacing: 30 cm.

### WEM climate panel

Composition: earth, straw and natural additives. Integral plastic and metal composite piping 16 × 2 mm. Glass-fibre fabric facing. Production: individual pressed. Heating elements for wall lining. Output: 85 W/m² at 35 °C, 170 W/m² at 45° (5 °C temperature difference between supply and return). Formats: 200, 160 or 80 × 62.5 × 2.5 cm. Subconstruction spacing: 60 cm, ceilings and roof inclines 30 cm, mounting threads embedded during production.

In addition to the above, a number of unfired panel-shaped products are produced by the brick industry: these are typically extruded perforated panels.

The subconstruction for clay panels must be sufficiently stable; the dimensions depend on the wall height and loads or wear it will be exposed to. As with other drywall constructions, appropriate constructions for mounting heavy items to the wall should be integrated into the subconstruction. Installations can be routed in the wall cavity. The spacing of the subconstruction, the fixing method for the panels and the number of fixing points per panel varies from product to product. Panel joins should ideally be arranged so that ends can be fixed to the subconstruction. Panel joins that occur midway between studs need to be fixed to a backing piece. The product manufacturers provide specific recommendations regarding suitable cutting tools. In most cases it is necessary to reinforce the joins between panels. The kind of reinforcement and its execution depends on the product.

Where clay panels are used as dry plaster panels stuck onto a flat surface, the subsurface must be stable, dry, clean and free of dust. The panels or large tiles are usually fixed in a bed of mortar or adhesive applied to the full surface. Different manufacturers recommend different adhesives. In most cases reinforcement fabric is applied to the full surface when plastering instead of just over the panel joins.

Subsequent surface treatment usually involves the application of a thin layer of plaster. It is important to be aware that some products are especially susceptible to swelling when wetted from one side. As a result, thicker layers of wet plasters can be problematic, especially when applied to extruded clay panels. It is important, therefore, to consult and adhere to the respective manufacturer's recommendations. In addition, some extruded panel products have a very smooth surface similar to unfired clay bricks. Additional measures to ensure a good plaster bond may be necessary.

Lightweight objects can be fixed to clay panels in different ways, for example with screws or cavity fixings. Consult the manufacturer's documentation for permissible loads.

## 6.3 Dry stacked walling

Heavyweight earth constructions that serve as a thermal storage buffer can be realised either in the form of wall infill or as a wall lining on the inner surfaces of walls. They are typically used to increase the mass of lightweight timber frame constructions to improve thermal comfort. In rooms with large areas of glazing, massive earth walls can also serve as passive solar energy collectors, reducing the risk of overheating. Due to the time-lag in the emission of stored heat, the indoor room temperature can be kept lower and more constant. As a result the thermal properties of timber constructions can be made more like those of massive constructions. This also contributes to conserving energy.

Thermal wall linings can be arranged along both internal as well as external walls. For internal walls one should bear in mind the not inconsiderable load of a massive wall lining. Where vertical walls do not align from floor to floor, additional structural measures may be required to distribute the extra load. When used for external walls, the thermal wall lining is not an alternative to insulation but rather an inward-facing wall lining. The thermal insulation should not be compromised.

Unlike masonry walling, the blocks are stacked and then clamped in place. This dry construction technique is comparatively quick to erect and reduces the moisture introduced into the construction. As the process of stacking blockwork is much quicker than laying blockwork it is also cheaper. Similarly, this comparatively straightforward dry construction method does not place such high demands on the block properties, and it is sufficient to use lower grade, less moisture-resistant Usage Class III blocks, which are generally cheaper (see chapter 8).

## 6.3.1 Dry stacked wall infill

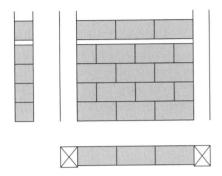

Fig. 6.4 : Dry stacked wall infill

Where earth blocks are stacked between vertical posts, some of them will need to be cut to fit between the posts. It is also possible to adapt the post spacing to match a multiple of block lengths, but it is important to take into account the larger dimensional tolerances of earth products, which can deviate more widely from standard brick dimension. The blocks should always be stacked in brick bond and clamped in place at not too large intervals by horizontal battens or sections of plank, fixed at their ends to the vertical posts. The posts and the clamping battens should be a few millimetres thinner than the block thickness so that subsequent facing boards presses the bricks firmly in place without leaving cavities. Alternatively, thicker posts can also be employed and the resulting cavity then used to route electrical cabling, for example.

## 6.3.2 Dry stacked wall lining

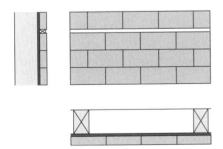

Fig. 6.5 : Dry stacked wall lining

Earth blocks can also be stacked on end as a wall lining in front of a wall. In such cases they should be held in place with a batten every three to four courses, that is every 37.5 to 50 cm. As with the wall infill, the battens should be slightly thinner than the thickness of the blocks. They should be secured at not too large intervals (seen horizontally) to the posts or timber particle board backing. The blocks should always be stacked in brick bond. Where heavy items will need to be attached to the wall, scantlings or suitable supporting constructions should be integrated into the construction beforehand. The blocks can be stacked in such a way as to leave space for vertical installation ducts, which if necessary can be lined with battens. Horizontal channels are best arranged just beneath the ceiling or alternatively, when the wall lining above is sufficiently supported, along the base of the wall. Electrical cables can be routed along recesses left between the stacked blocks or in chases cut into the wall lining. Recessed electrical socket boxes can be inserted as usual and, where necessary, fixed to the backing particle board. The completed stacked wall lining is then typically covered with drywall plasterboard, ideally clay plasterboard panels (see section 6.2). The distance between the posts or the clamping battens should be chosen to match the dimensions and necessary fixing intervals of the subsequent plasterboard panels.

If more than a very thin layer of earth plaster is to be applied, the drying should carefully controlled and monitored. Towards the end of the building process, the necessary air exchange rate in a timber construction can only be achieved by active or intentional means, as the optimised building envelopes of modern low-energy or passive-energy houses drastically reduce the general level of natural ventilation. If a wet plaster is intended, a plaster backing, such as reed matting, must first be applied to the stacked wall lining. The mat should be fixed to the posts and / or the clamping battens making sure not to exceed the maximum fixing intervals. The plaster base must be firmly fixed and should not flex appreciably.

## 6.4 Ceiling overlays and ceiling and roof infill

Earth building materials are used to add mass to ceilings to reduce sound transmission or retain warmth. In roofs they are used primarily to counteract overheating in summer.

In ceilings, layers of earth can be arranged either in a layer above the floorboards or between the ceiling joists. *Ceiling overlays* are most commonly used in ceilings where the full cross-section of the beams should remain visible on the underside. In the case of *ceiling infill*, a portion of the beams can be visible on the underside but more often the entire underside is clad. This approach corresponds to the historical technique described in section 10.4 except that dry earth building materials are used for the infill material.

Fig. 6.6 : Alternative ceiling constructions

In terms of reducing sound transmission, ceiling overlays are advantageous, in particular compared with ceiling infill with partially visible beams on the underside. Ceiling infill, on the other hand, makes it possible to create thinner ceiling constructions.

Before installing ceiling overlays or ceiling infill it is essential to establish that the ceiling is able to support the extra load. Before the finished floor surface or floorboards are installed, the residual moisture content must be checked in case the earth materials absorbed moisture during transport or the building process.

## Ceiling overlays

Heavy earth bricks or blocks of Usage Class III are usually used for ceiling overlays. In most cases, a layer of dust-preventing building paper is needed. This continuous, uninterrupted layer serves to lessen airborne noise transmission and each strip should, therefore, sufficiently overlap the next. At junctions with walls or parts of the building that penetrate the floor, the paper should be turned up and stuck in place. In timber constructions, building papers made of reinforced card or paper are most commonly used. If plastic foils are to be used, it is important to consider their effect as a vapour barrier and to ensure that condensation water will not collect on wood surfaces.

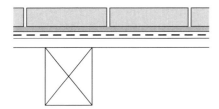

Fig. 6.7 : Cross-section with ceiling overlay

The impact sound insulation of the ceiling can be improved by inserting a layer of felt between the building paper and the earth bricks. Bedding the bricks in mortar, in a manner similar to the laying of concrete paving blocks in bitumen, is difficult because green unfired earth bricks are susceptible to moisture uptake. The bricks are usually laid in brick bond which makes it easy to keep them aligned. It is not necessary to lay them with open joins that are then subsequently filled with a brushed-in mass.

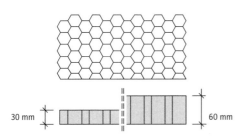

Fig. 6.8 : Honeycomb floor element

Dry ceiling overlays can also be realised using loose fill material. The dry granulate can be filled in the traditional manner between battens and levelled off. Another effective method is to fill in 30 or 60 mm high hollow *honeycomb elements*. These help to stop the fill material from shifting around as a result of vibrations or pressure during later use. After installation all loose fill materials must be protected against moisture ingress, mechanical loads and soiling. With dry fill materials an additional protective layer of lining paper can be necessary to stop dust getting into the room above.

Ceiling infill

Ceiling infill consists of supporting battens, timber boarding and the earth fill material. Sufficiently large battens are nailed to the side faces of the ceiling joists, for example along the lower edge. The timber boarding is laid between them and can consist of timber boards or composite wood-based panels. They are typically fixed on either side to the bearing battens or at least held in place so that they do not slide about during the construction phase. Before a loose-fill material is applied, a layer of building paper should be applied. The density of the fill material can be varied to meet structural requirements by mixing in lightweight additives. It is also possible to fill a part of the cavity with a lightweight insulation material to dampen noise transmission.

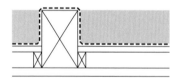

Fig. 6.9 : Cross-section: ceiling infill

The earth infill can also be realised using earth blocks of Usage Class III or another usage class. This is easiest when the distance between the ceiling joists can be adjusted to correspond to a multiple of the brick dimensions. It is important to be aware that earth products can deviate more significantly from the standard dimensions compared with fired bricks. Gaps in the layer of bricks can be filled with a loose fill material. A cavity above the layer of bricks is best filled with lightweight insulation. The underlay for the finished floor can be realised in different ways. Typical variants include preformed hollow fibre insulation board with battens or the use of dry screed flooring boards.

Fig. 6.10 : Cross-section with preformed hollow fibre insulation board (left) & dry screed flooring boards (right)

Roof infill

Roof infill between the rafters in the roof spaces is realised in much the same way as ceiling infill, except that the earth infill material has to be held in place to prevent it sliding. Hollow block slabs slung between the rafters can also be used.

## 6.5 Material and building element properties

### 6.5.1 Mechanical properties

Earth block wall linings have no stiffening effect. This applies likewise for clay plasterboarding. Values for permissible fixing loads and pull-out resistance are not available for clay plasterboards. With the exception of very light loads, items should be fixed to the underlying subconstruction.

Table 6.1 : Weight per unit area of vertical or horizontal earth block linings

| Block bulk density $kg/m^3$ | t = 5.2 mm (DF) $kg/m^2$ | t = 7.1 mm (NF) $kg/m^2$ | t = 11.3 mm (2DF) $kg/m^2$ |
|---|---|---|---|
| Blocks 1300 | 67.6 | 92.3 | 146.9 |
| Blocks 1500 | 78.0 | 106.5 | 169.5 |
| Blocks 1900 | 98.8 | 134.9 | 214.7 |

Table 6.2 : Weight per unit area of clay plasterboard and clay plaster panels

| Panel/plaster bulk density $kg/m^3$ | Thickness mm | Weight per unit area $kg/m^2$ |
|---|---|---|
| Panels 700 | 16 | 11.2 |
| Panels 700 | 20 | 14.0 |
| Panels 700 | 25 | 17.5 |
| Panels 1200 | 14 | 16.8 |
| Panels 1200 | 22 | 26.4 |
| Panels 1440 | 25 | 36.0 |
| Panels 1600 | 25 | 40.0 |
| Plaster 1600 | 5 | 8.0 |
| Plaster 1600 | 10 | 16.0 |

Table 6.3 : Compressive strength of earth blocks of usage class III

|  | Compressive strength, N/mm² |
|---|---|
| from | 1 |
| mean | 3 |
| to | 10 |

## 6.5.2 Thermal insulation, thermal capacity and water vapour diffusion resistance

Table 6.4 : λ-value, specific heat capacity c and μ-value of dry earth building materials

| Bulk density kg/m³ | λ-value W m/K | Specific heat capacity c kJ/kgK | μ-value |
|---|---|---|---|
| Blocks 1300 | 0.53 | 1.0 | 5/10 |
| Blocks 1500 | 0.66 | 1.0 | 5/10 |
| Blocks 1900 | 1.00 | 1.0 | 5/10 |
| Panels 700 | 0.21 | 1.1-1.3 | 5/10 |
| Panels 1200 | 0.47 | 1.0-1.1 | 5/10 |
| Panels 1400 | 0.59 | 1.0-1.1 | 5/10 |
| Panels 1600 | 0.73 | 1.0 | 5/10 |

## 6.5.3 Sound insulation

The sound insulation of dry earth building constructions has been determined in a test series (DIN EN 20 140-3 / ISO 140-3) undertaken with CLAYTEC clay plasterboard.

Table 6.5 : Test values $R_{W,P}$ and calculated values $R_{W,P}$ of dry earth building constructions

| Structure of the test specimen [1] | Test value $R_{W,P}$ dB | Calculated value $R_{W,R}$ dB |
|---|---|---|
| 3 mm clay plaster<br>25 mm CLAYTEC clay plasterboard, m' = 17 kg/m² | 36 | 34 |
| 3 mm clay plaster<br>25 mm CLAYTEC clay plasterboard, m' = 17 kg/m²<br>60 mm air cavity<br>25 mm CLAYTEC clay plasterboard, m' = 17 kg/m²<br>3 mm clay plaster | 47 | 45 |

Table 6.5 : continued

| Structure of the test specimen [1] | Test value $R_{W,P}$ dB | Calculated value $R_{W,R}$ dB |
|---|---|---|
| 3 mm clay plaster<br>25 mm CLAYTEC clay plasterboard, m' = 17 kg/m²<br>75 mm unfired NF (24 × 11.5 × 7.4), m = 3.7 kg/brick<br>25 mm CLAYTEC clay plasterboard, m' = 17 kg/m²<br>3 mm clay plaster | 48 | 46 |
| 3 mm clay plaster<br>25 mm CLAYTEC clay plasterboard, m' = 17 kg/m²<br>80 mm HOMATHERM insulation board<br>25 mm CLAYTEC clay plasterboard, m' = 17 kg/m²<br>3 mm clay plaster | 53 | 51 |
| 3 mm clay plaster<br>25 mm CLAYTEC clay plasterboard, m' = 17 kg/m²<br>70 mm sheep's wool, loosely-inserted<br>25 mm CLAYTEC clay plasterboard, m' = 17 kg/m²<br>3 mm clay plaster | 56 | 54 |
| 3 mm clay plaster<br>25 mm CLAYTEC clay plasterboard, m' = 17 kg/m²<br>56 mm unfired NF (24 × 11.5 × 5.6), m= 2.95 kg/brick<br>14 mm air cavity<br>15 mm OSB-board (oriented strand board) | 43 | 41 |
| 3 mm clay plaster<br>25 mm CLAYTEC clay plasterboard, m'= 17 kg/m²<br>56 mm unfired NF (24 × 11.5 × 5.6), m= 2.95 kg/brick<br>15 mm OSB-board (oriented strand board) | 43 | 41 |
| 3 mm clay plaster<br>25 mm CLAYTEC clay plasterboard, m'= 17 kg/m²<br>80 mm HOMATHERM insulation board<br>15 mm OSB-board (oriented strand board) | 46 | 44 |

[1] rows 1-2: size of test specimen 4120 × 2180 mm, rows 3-8: size of test specimen 2015 × 1015 mm, rows 1-5 & 8: subconstruction, 4/6 cm battens, rows 6-7: subconstruction, 4/4 cm battens.

From these values the following sound reduction improvement and sound reduction index values $R_W$ can be derived (calculations by the SWA-Institut in Aachen):

Table 6.6 : Sound reduction improvement values of dry earth wall linings

| Internal dimension cm | No infill dB | Earth blocks dB | Insulation dB | Insulating wool dB |
|---|---|---|---|---|
| 6 | 8 | 8 (DF) | 21 | 25 |
| 8 | 10 | 11 (NF) | 23 | 27 |
| 10 | 12 | 15 (2DF) | 25 | 29 |

*Derived values calculated by the SWA-Institut, Aachen.*

Table 6.7 : Sound reduction index values for partitioning walls in earth dry wall construction

| Internal dimension cm | No infill dB | Earth blocks dB | Insulation dB | Insulating wool dB |
|---|---|---|---|---|
| 6 | 47 | 46 (DF) | 51 | 54 |
| 8 | 49 | 48 (NF) | 53 | 56 |

*Derived values calculated by the SWA-Institut, Aachen.*

### 6.5.4 Fire performance

Building material class

Clay plasterboard panels *without* organic additives are classified in the currently valid DIN 4102-4 (March 1994) by analogy as *non-combustible*.

The fire performance of clay plasterboard panels has been determined in tests (DIN 4102-4) undertaken with CLAYTEC clay plasterboard.

Table 6.8 : Fire-resistance rating of walls with clay plasterboarding

| Structure of the test specimen [1] | Fire-resistance rating |
|---|---|
| 3 mm clay plaster<br>25 mm CLAYTEC clay plasterboard, m' = 17 kg/m² [2]<br>60 mm air cavity<br>15 mm OSB board, tongue and groove [3] | F 30 |
| 3 mm clay plaster<br>25 mm CLAYTEC clay plasterboard, m' = 17 kg/m² [2]<br>60 mm HOMATHERM insulation board<br>15 mm OSB board, tongue and groove [3] | F 30 |
| 3 mm clay plaster<br>25 mm CLAYTEC clay plasterboard, m' = 17 kg/m² [2]<br>60 mm Luftschicht<br>18 mm GUTEX MULTIPLEX N, tongue and groove [4] | F 30 |
| 3 mm clay plaster<br>25 mm CLAYTEC clay plasterboard, m' = 17 kg/m² [2]<br>60 mm HOMATHERM insulation board<br>18 mm GUTEX MULTIPLEX N, tongue and groove [4] | F 30 |

[1] Listed in order from flame-exposed surface to reverse surface.
[2] Fixed on 60 × 40 mm battens (50 cm centres) with 5 × 70 mm SPAX screws and D 20 mm washers at 30 cm intervals.
[3] Fixed with 5 × 50 mm SPAX screws, 4 per stud and board.
[4] Fixed with 5 × 50 mm SPAX screws at 35 cm intervals.

## Fire-resistance rating

The fire-resistance classes of walls and ceilings with clay plasterboarding have been determined in a test series (DIN 4102-4) undertaken with CLAYTEC clay plasterboard.

Table 6.9 : Fire-resistance rating of ceilings with clay plasterboarding

| Structure of the test specimen [1] | Fire-resistance rating |
|---|---|
| 3 mm  clay plaster<br>25 mm  CLAYTEC clay plasterboard, m' = 17 kg/m² [2]<br>40 mm  air cavity [3]<br>160 mm  air cavity [4]<br>20 mm  rough tongue and groove boarding [5] | F 30 |
| 3 mm  clay plaster<br>25 mm  CLAYTEC clay plasterboard, m' = 17 kg/m² [2]<br>40 mm  air cavity [3]<br>60 mm  HOMATHERM insulation board [4]<br>100 mm  air cavity<br>20 mm  rough tongue and groove boarding [5] | F 30 |

[1] Listed in order from flame-exposed surface to reverse surface.
[2] Fixed on 40 × 60 mm battens (36 cm centres) with 5 × 70 mm SPAX screws
  and D 20 mm washers at 30 cm intervals.
[3] Battens fixed to the beams with 5 × 70 mm SPAX screws, 2 per batten and beam.
[4] Beams, 160 × 200 mm (90 cm centres).
[5] Fixed with 65 mm panel pins, 2 per board and beam.

# 7 Internal insulation with earth materials

## 7.1 Internal insulation in general

### 7.1.1 Introduction and background

Internal insulation describes layers of insulation ($\lambda \leq 0.1$ W/mK) or thermally insulating building materials that are applied to the internal surface of external walls. This technique is generally used where it is not possible to insulate the external surface of a building, for example when the façade of a building or face of the building's frame structure must remain visible. A further reason for its use is when only sections of a façade need insulating (e.g. individual rooms or storeys). Internal insulation is also employed to decouple the thermal mass of an external wall from the heating of the room. This makes it easier to heat the room and to bring its thermal characteristics more in line with modern patterns of use. This can, however, create problems in summer where there is a need to protection against excessive heat.

Compared with insulation located on the outer surface of external walls, internal insulation can be problematic in terms of condensation. Condensation occurs when the water vapour content of the air reaches saturation (100 % relative humidity). In winter, warm air from the interior, which holds much more moisture than the cold air outside, diffuses through the wall from the heated side to the colder side. As it passes through the wall the air cools down progressively. Because cold air can hold less water vapour than warm air, the saturation level is reached at some point in the cross-section of the wall. The water vapour condenses into capillary conducted water. This effect is heightened by compact and sealing layers on the outer surface which act as a barrier containing the flow of moisture within the wall.

A degree of so-called interstitial condensation within the building construction is tolerable, but it should not exceed certain limits (table 7.1) and must also be able to dry out during warm times of the year. If this is not the case the insulation will at some point no longer fulfil its function.

It is theoretically possible to place a vapour barrier on the inner surface to prevent air and airborne moisture from passing into the construction. However, this approach can have serious consequences, especially where façades are exposed to the weather: a vapour barrier also acts as a barrier for moisture contained within the construction, such as rainwater from driving rain, preventing it from escaping and drying. As the amount of moisture in a construction caused by exposure to rain is far greater than that caused by interstitial condensation, priority should be given to ensuring that the construction can dry out. This applies especially for timber-frame façades where rainwater penetrates the gap between timber members and the panel infill material. Another problem associated with vapour barriers is that they have to be installed correctly, so that there are no holes or leaks, not just in the surface of the wall but particularly at the edges and junctions. For timber constructions this is hard to achieve flawlessly and only at considerable cost.

Aside from the problem of interstitial condensation, internal insulation can also be problematic for façades exposed to the weather: the thicker the insulation, the more it prevents the wall from heating up. Moisture contained within a warm wall dries out more quickly than in a cold wall. In certain circumstances timber members in an internally insulated wall construction can be subject to prolonged exposure to moisture for longer than before the wall was insulated. A construction that may have survived for decades without defects may start exhibiting problems after the installation of internal insulation.

### 7.1.2 Requirements of building materials for internal insulation

Given the problem described above, all of the internal insulation variants described here do not function as a vapour barrier.

Capillary conductive insulation materials should be used so that condensation can be dissipated throughout the material rather than concentrate in one place. The moisture is conducted through capillary action inwards and outwards towards the surface where it can then evaporate.

Furthermore, materials for internal insulation should have a good sorption capacity: they should be able to retain moisture through adsorption within the material in order to reduce the degree of moisture in the vapour before the saturation level is reached.

Similarly, the installation technique should ensure continuous capillary contact between the materials without voids, gaps or air cavities. For this reason soft and malleable insulation materials are advantageous, or alternatively insulation boards bedded firmly in mortar.

### 7.1.3 Suitability of earth building materials for internal insulation

Earth building materials are capillary conductive and exhibit good sorption characteristics. When wet they are malleable and can be used to easily create void-free constructions. As such they fulfil the above conditions well.

A limitation, however, is that thermal insulation materials are defined as having a *specific thermal conductivity* of $\lambda \leq 0.1$ W/mK. The thermal conductivity of light earth materials is not as good with thermal conductivity values of $\lambda \geq 0.12$ W/mK and greater. Due to the particularly sensitive nature of the problem of internal insulation, this can, however, be advantageous. The occurrence of interstitial condensation depends considerably on the thermal conductivity of the wall. In the case of light earth internal insulation, its lower conductivity means that less or even no moisture occurs within the wall.

### 7.1.4 Dimensioning internal insulation

The aim when dimensioning internal insulation is to achieve the maximum insulation effect while simultaneously avoiding the risk of excessive moisture in the structure. The objective is to find an optimum level that satisfies these two contradictory requirements. A further requirement is to minimise the amount of space it takes away from the interior. If wall heating is to be installed on the internal surface of external walls, the thermal conductivity of the wall should not exceed the U-value of 0.45 W/m²K according to the *German National Association for Surface Heating and Cooling (Bundesverband Flächenheizungen und Flächenkühlungen)*. This value cannot always be achieved, especially when renovating timber-frame constructions.

The thermal insulation effect is calculated by determining the U-Value as described in DIN 4108.

There are a number of different methods for calculating and demonstrating the occurrence of interstitial condensation.

The methods described in the DIN include:
- The *Glaser* method, which uses fixed climatic data without taking into account the capillary conductivity and sorption characteristics.
- EN ISO 13788, similar to the above but using more realistic climatic data and taking into account possible interstitial condensation in several layers.

Other recognised procedures include:
- The *Jenisch* method, which is similar to the *Glaser* method but uses real climatic data.

- COND: a program for the hygrothermic assessment of constructions. It too uses a similar basis to Glaser but extends it by taking into account moisture transport within the construction.
- DELPHIN, WUFI: simulation programs for calculating combined heat, moisture, air and salt transport processes in porous building materials.

DIN 4108 Part 3 prescribes calculation according to the *Glaser* method and the fulfilment of particular requirements and boundary conditions for condensation occurrence. This requires an at least approximate assessment of the material properties of the existing building.

More precise analyses using more detailed techniques such as *COND* and computer-aided simulations such as *WUFI* and *DELPHIN* are permitted. Characteristic values for

Table 7.1: Requirements according to the current state of the art: DIN 4108-3 2002-02; WTA technical information sheet 8 – timber-frame constructions; Lehmbau Regeln 2009

|  | Requirement | Source |
|---|---|---|
| Minimum thermal insulation | $R_{ges} \geq 1.2$ m²K/W | DIN 4108 |
| General prevention of condensation [1] | Drying out in summer | DIN 4108 |
|  | $\leq 1000$ g | DIN 4108 |
|  | Wetting of wood $\leq 5\%$ | DIN 4108 |
|  | Wetting of composite wood material $\leq 3\%$ | DIN 4108 |
| Ability of wall constructions to dry out, surfaces exposed to the weather, driving rain load class I [2] | $\leq 500$ g | WTA |
|  | Avoidance of vapour barriers | WTA |
|  | Capillary conductive insulation material | WTA |
|  | Good interlayer contact for continuous capillary conduction | WTA |
|  | Leak-free and void-free construction | WTA |
|  | $S_{di}$ 0.5 – 2.0 m | WTA |
|  | $R_i \leq 0.8$ m²K/W | WTA |
| Reduction of construction moisture content during installation | Cavity filling and levelling layer with earth mortar in wet consistency, D $\leq 3$ cm | LR |
|  | Light earth in wet consistency, D $\leq 15$ cm [3] | LR |

[1] No further proof necessary where $R_i \leq 1.0$ m²K/W and $S_{di} \geq 0.5$ m.
[2] For driving rain load classes II and III, a form of façade cladding is necessary. Classification according to driving rain load class should take into account the respective actual orientation and exposure to weather.
[3] D $\leq 20$ cm is permissible in the case of external walls made of vapour-permeable and capillary conductive building materials (straw-clay, bricks with a bulk density $\leq 1600$ kg/m³).

use in calculating the simulation are as yet not available for all earth building materials, and likewise for many materials in existing historical buildings.

Independent of the issue of insulation, the need to protect against driving rain must be taken into account. According to the *Technical Work Group of the German Building Renovation and Conservation Association (WTA)*, as well as other professionals in the field, in regions with high driving rain loads (level II and III according to DIN 4108 Part 3) timber frame constructions should only be left exposed on sheltered façades, that is elevations not exposed to the weather or in densely built-up settlements. The actual exposure of a façade to wind and rain can only be assessed in each individual case [WTA, 2003].

For the internal insulation of timber-frame walls exposed to wind and rain, the WTA propose that the insulation including any internal finishes should be limited to a thermal conductivity of 0.8 m²K/W. At the same time the diffusion equivalent air layer thickness or $S_d$-value of these layers total between 0.5 and 2.0 m.

In addition, one should also take into account the need to limit the amount of moisture introduced during construction as well as the need to ensure that the construction can dry safely.

The internal insulation methods described here that do not feature vapour barriers are generally not sufficient to fulfil other standards such as the *Energy Conservation Regulations (EnEV)*. For buildings with façades that cannot be insulated externally, for example because they are listed buildings, thermal insulation is secondary to preservation of the construction. § 24 and § 25 of the EnEV detail permitted exemptions and exceptions to the rules given elsewhere in the EnEV.

## 7.1.5 Preparing existing walls for internal insulation

The base of the existing external wall must be free of rising damp or other sources of moisture. It is, however, not always possible to install a horizontal damp proof course or external seal. A remedy can be to use calcium silicate panels on the internal side of the wall. These direct moisture via capillary action to the surface of the wall where it can evaporate. They also function as an insulator with a thermal conductivity of 0.65 W/mK.

The use of sealants on the inner surface, such as with bitumen or a vapour-proof insulation material such as foam glass insulation boards, should be viewed with caution: they can result in the moisture simply rising higher than before in the wall.

In any case, it is advisable to consult an expert with appropriate knowledge and experience in the repair of walls with rising damp.

Salt contamination is a further problem. Soluble salts are contained in the moisture and transported in the wall. Through their hygroscopic action they can themselves be the cause of wet surfaces, even after correctly implemented sealing. Crumbling plaster and traces of earlier repairs to the wall can be an indication of salt contamination. The walls of old stables and buildings for livestock can be particularly problematic due to salts from manure. It is imperative to investigate this aspect before embarking on converting such buildings.

Any non-permeable wall coverings and coatings that could interrupt capillary action must be removed. These include coatings designed to protect against splash water (tiles, oil paint, varnish) and decorative wall coverings and coatings (vinyl wallpaper, dispersion paints). Similarly, coatings made of organic material that could disintegrate when exposed to moisture (paper wallpaper and coverings) should also be removed.

The existing internal plaster is in most cases already sufficiently vapour permeable. Multiple layers of paste can however form a barrier. A simple means of checking the permeability of a surface is to wet it with water. The water is applied with a brush. If the water is absorbed by the wall within a few minutes, the wall has good absorbency characteristics, which in turn is an indication of good vapour permeability.

Installations and electrical cables are in principle possible in both the layer of insulation as well as bonding mass. It is important, however, to ensure that the water pipes will not be subject to frost and that condensation will not form on the installations. The minimum level of insulation should also apply for the installations. It is also important to consider the possible moisture effect of installations and cables during the drying phase of the inner wall lining. For this reason, installations are in practice very often arranged in front of the insulation or against the internal walls.

### 7.1.6 Junctions with walls and ceilings, and window and door reveals

Where walls or ceiling join the internal insulation, it is recommended to insulate the surfaces directly adjoining the external wall. The objective is to minimise energy loss through the cross-wall or ceiling and prevent critical cooling of the wall surface. Because this is complex to realise and the sections of insulation or panel edges look unsightly in the room, it is worth noting the following:

The danger of surface condensation and mould formation is not usually a problem with historical building materials such as wood and earth because the thermal conductivity is not so bad that the surfaces could cool to such critical temperatures. The same applies for ceramic bricks with a bulk density of $\leq 1600$ kg/m$^3$.

Many other constructions, such as walls made of brick with a bulk density of $\leq 1800$ kg/m$^3$ can be unproblematic. The relative humidity in rooms of centrally heated old buildings in winter is more likely to be in the range of 40 % rather than 50 %. Calculations show that for a relative humidity of 40 %, surface temperatures are generally non-critical.

In addition, building materials and surfaces with good sorption capacity can counteract the effect of high indoor air humidity through their buffering effect. Absorbent earth plasters minimise the risk as surface condensation water is dissipated, preventing the occurrence of concentrated areas of moisture.

Walls and ceilings made of dense, highly thermal conductive materials such as concrete should be insulated along the junction to the external wall. Special care should be taken when mould formation was already a problem before beginning the insulation measures. Similarly, caution is advisable where many measures have been taken to improve the air tightness of the room with the aim of minimising air exchange in winter or in the case of poorly ventilated rooms with continually high humidity levels.

Window and door reveals should where possible be additionally insulated. Even a minimal thickness of 20 mm is effective at increasing the surface temperature. The necessary space can be gained by removing the old plaster. New bonding compound and plaster layers can be executed thinner than normal if required. Air draughts around the window frame should be prevented (air tightness). Special care should be taken when mould formation was already a problem prior to beginning the insulation measures. In critical cases, mineral-based reveal insulation panels are recommended.

### 7.1.7 Bearings for timber joists

To properly assess the bearings of timber joists a specialist should be consulted. This is especially important where the walls are massive and a thick layer of internal insulation is planned: on the one hand, water vapour can condensate on the now cooler surface of the timber joists (due to the new insulation) if room-temperature air is able to reach the joists, for example through a leak. On the other hand, the external wall is more or less decoupled from the room heating by the insulation. This can extend the drying time after moisture ingress through rain and the ends of timber joists may be subjected to prolonged contact with moisture.

To begin with the existing situation of the timber joists where they rest on the wall should be examined. If damaged ends of timber joists need replacing, this should be undertaken with the necessary care and attention. In particular timber members should not rest directly on or in brick masonry. The beams should rest on a separating layer (for example bituminous paper). Similarly, any other surfaces of the wood should not directly touch the masonry. There should always be a gap. On the room-ward side, the junction should wherever possible be airtight.

If the wood is intact, the danger of air reaching it directly, for example through joints between floorboards, should be examined. Floors with largely closed joints and no visible direct connection to the external face of the façade do not need to be taken up. Likewise, the removal of intact wall and ceiling plaster that already forms an airtight seal should be considered carefully: the "perfect" detail is often only supposedly better than the existing situation. If the decision is taken to open the ceiling above, the gaps between masonry and the timber joists should be plugged with a suitable insulating wool. The air-tightness of the junction should be effected as best possible, for example by smearing with earth plaster mortar. Sealing with adhesive tape is rarely very effective where old buildings are concerned. The mortar also has the advantage of being capillary conductive.

All of the comments above apply equally for the bearings of timber joist ceilings in timber frame constructions. These are generally much less complex as they are either exposed and therefore able to dry out quickly or encased in plaster that has good capillary conductivity.

### 7.1.8 Minimising air leakages with internal plastering

All of the internal wall linings described below should be free of voids, cavities and leak-proof. The bonding compound of insulation boards and layers of internal plaster can also be used to minimise leaks (section 7.4). Continuous, uninterrupted layers of plaster can act as an airtight layer (chapter 4). In the case of timber-frame buildings it is not always possible to ensure complete air tightness. In such cases they should at least be minimised by skilled, careful workmanship. Gaps should be carefully filled, holes sealed with earth mortar.

## 7.2 Light earth wall linings (wet construction)

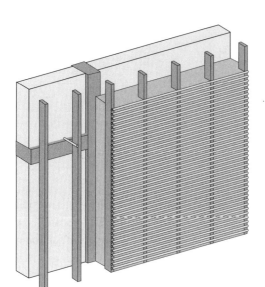

Fig. 7.1 : Construction principle: light earth wall linings

### 7.2.1 Description of the internal insulation

Light earth wall linings consist of a framework of vertical battens and light earth fill material. The vertical battens serve as a supporting structure for formwork. The void between the formwork and the existing external wall is filled with a naturally-moist to malleable mixture of earth and lightweight aggregates. In most cases light earth wall linings are plastered once dry; sometimes they are clad with boards.

### 7.2.2 Constructing light earth walls

When deciding on the thickness of internal wall linings made of moist light earth, it is important that the construction can dry out rapidly. The *Lehmbau Regeln* limits the possible thickness to a mean thickness of ≤ 15 cm. If the external wall consists of materials with good capillary conductivity, such as straw-clay or brickwork (bulk density ≤ 1600 kg/m³), a thickness of ≤ 20 cm is permissible. The limitation of the thickness has two reasons: to prevent the organic constituents in the mixture from decomposing and to lessen the exposure of the existing structure, and structural timber elements in particular, to excessive moisture during construction.

Before erecting the framework of battens, the construction and junction details must be decided on. Massive constructions are usually sufficiently planar and upright; for timber-frame buildings the sill beam often serves as a suitably linear reference line for constructing the framework. The battens can be fixed at the top with a crossbar fixed to the ceiling joists.

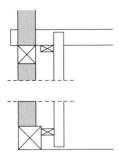

Fig. 7.2 : Upper and lower fixing detail of the battens of the subconstruction

Window openings are typically given splayed reveals so that as much light as possible can enter the room. Above the window opening, a solid lintel made of a scantling is needed to support the weight of the wall lining above it. In the corners of the room, two vertical battens are required to receive the formwork of both wall planes where they meet at the corner.

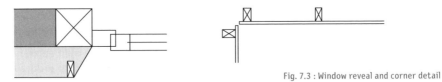

Fig. 7.3 : Window reveal and corner detail

Planks or other timber elements can be integrated into the batten construction to act as a secure mounting for heavier items such as radiators, wall-hung cupboards and so on.

The vertical battens must be firmly fixed at the top and the bottom to resist tension. An additional means of anchorage should also be provided in the middle to prevent the batten buckling when the light earth is forced in behind it. This can be achieved with strong wire or perforated metal strip of the kind used for wall ties.

Vertical installation channels can be achieved by placing squared timbers in the construction that are then removed once the construction has been filled. Another possibility is to mount two vertical battens a short distance apart and not to fill the space between them. Horizontal installation channels can be created, for example, at the base of the wall by making the vertical battens spring back at the base.

*Reusable formwork* is removed once the wall lining has been filled. It should be made of sufficiently stable and moisture resistant 24 mm thick shuttering planks or boards of a composite wood material that are fixed to the vertical battens with screws and washers. If the joins do not meet at a batten, they should be fastened with straps to prevent them shifting out of place. The formwork can usually be removed immediately after backfilling with light earth, or after a short drying time in the case of expanded clay light earth.

*Permanent formwork* remains in place after use. It should be stable but also open and air-permeable so that it does not impede the drying process. Rolls of reed plaster lath with a stem density of ≤ 70 stems/m are suitable (not to be confused with reed insulation boards). Another alternative is to use narrow boards or battens set sufficiently apart from one another to allow drying.

Reed plaster laths are stapled to the vertical battens with staples of at least 25 mm. Rather than using the wire of the mesh, an extra 1.2 to 1.5 mm thick additional wire should be used. Adjacent sections of rolled lath should meet at the vertical battens. The lath is fixed at the bottom of the battens and rolled successively upwards and fixed. The stems lie horizontally and serve as a plaster lath for subsequent plastering. Where timber plaster laths are used they should be screwed to the vertical battens.

At the start of work, only a short amount of wall should be closed off with shuttering. As the process of backfilling the void becomes more familiar, the height of a layer can be increased to as much as 40 cm. The larger the height of formwork fixed in advance, the harder it becomes to properly fill the area behind it. Formwork made of reed plaster lath and timber lath have the advantage that the height can be chosen freely according to the conditions on site. A further advantage is that it is easy to ensure that the fill material is well compacted as gaps are immediately visible after filling in the mixture.

The light earth material is straightforward to transport on the building site. Spades or forks are commonly used to fill the cavity. Some mixtures are also suitable for pumping. The material should be filled so that a continuous mass results free of appreciable gaps or voids (several cm in diameter). For this reason it is recommended not to fill in too much material at once and to gently compact the mixture with a section of wooden post or batten. The material should be firmly packed but not rammed! The intention is to create a homogenous void-free mass of the required material density.

To ease working, the light earth wall lining can be extended upwards into the region of the timber joists. It does however require that the existing historical building fabric is removed to gain access to the area. This may not be necessary as old ceiling fill material rarely represents a cold bridge and they are often not dissimilar to light earth. The beams may even have low thermal conductivity.

At the uppermost section of the wall lining, the formwork is extended just far enough that one can still fill material from above. The final section is then closed off with the fill material from the side.

### 7.2.3 Types of light earth and specific aspects of their construction

Light earth is defined as having a bulk density of ≤ 1200 k/m³. One differentiates between organic and mineral light earth according to the type of additive. Mixtures of the two are also possible. A variety of different products are available on the market and are usually delivered ready to use. Organic light earth materials have a limited shelf life and they should be installed as soon as possible without long storage times.

In addition to the common pre-mixed variants described below, other kinds of light earth can be pre-mixed or mixed on site with a variety of different additives. These include straw light earth, which is no longer used in appreciable quantities. Its manufacture and application is described by Franz Volhard [Volhard, 1995].

#### Wood-chip light earth

Wood-chip light earth, as the name suggests, is a mixture of wood chips and earth. The lambda-value for a mixture with a bulk density of 600 kg/m³ is 0.17 W/mK according to DIN 4108. The material has a particulate, slightly lumpy structure but, when properly installed, is not subject to settlement. There is no danger of the wood chips rotting, either during the drying period or under normal exposure to moisture resulting from driving rain or interstitial condensation. The battens can be made of non-impregnated roofing battens. The horizontal spacing of the vertical battens should be around 30 - 35 cm. To ensure the battens do not bow, they should be fixed at the top and bottom as well as at half-height. Reed plaster lath is typically used as formwork.

#### Expanded clay light earth

This material consists of earth mixed with fired and expanded, porous clay particles. The granularity of the expanded clay typically lies between 4 - 8 or 8 - 16 mm, the bulk density of the light earth between 600 and ca. 700 kg/m³. The insulation value of the mixtures lies between 0.17 and 0.21 W/mK according to DIN 4108. The batten construction should be more stable than that used for wood-chip light earth wall linings as the spherical granularity of the additive provides little to no inner cohesion. Vertical ties should also be placed closer together than with other kinds of light earth. It is important to ensure that the additives do not become wedged behind the vertical battens but that they are held by the binding effect of the earth. Re-usable shuttering planks or boards are used as formwork. The filling and compaction of the mixture must be undertaken with care as the light earth surface will need to be sufficiently firm and homogenous to serve as a plaster base for subsequent plastering. The shuttering boards should be left in place for several days (1–6) after installation so that the material can harden sufficiently to stay in place. The shuttering must then be removed to allow the drying process to begin. It is also possible to use permanent formwork made of a sufficiently dense reed plaster lath.

Foamed glass light earth

Foamed glass light earth consists of earth, straw and foamed glass with a granularity of up to 4 mm. Foamed glass is exceptionally lightweight with a bulk density of 460 kg/m³. The thermal conductivity of this mixture is given as 0.114 W/mK.

Light earth with kieselguhr

This material, which consists of straw and granular aggregates is not strictly an earth building material as defined in the *Lehmbau Regeln* as the material cures and cannot be replasticised through the addition of water. The thermal conductivity of the mixture is 0.08 W/mK almost two times better than that of other light earth variants. The material exhibits high capillary conductivity and is therefore well suited for the internal insulation of timber-frame constructions. 40/60 mm or 40/80 mm battens arranged at ca. 60 cm centres are recommended for the supporting construction. An open formwork of thin boards usually serves as shuttering onto which plaster lath is fixed.

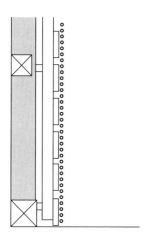

Fig. 7.4 : Batten framework, an open formwork of thin boards with additional reed plaster lath over a wall lining of light earth and kieselguhr

### 7.2.4 Building duration and drying

Light earth construction works are mostly undertaken during the warm summer months. They are, however, possible during winter in temperature-controlled or heated building sites where the high moisture capacity of cold-dry winter air can be advantageous for the drying process. As a rule, the building site should be permanently ventilated during *the entire drying period* and not just for a limited number of hours during the day. If this is not possible, a reliable means of forced drying should be organised. Wall linings that are in the process of drying may not be covered with building materials or other coverings. Temporary mould formation cannot be excluded with any of the aforementioned moist variants of light earth as light earth is chemically neutral, i.e. not fungicidal. After the first indication of mould formation, effective forced drying should *immediately* be implemented. The notes concerning the *drying of earth plasters* apply for the most part here too (see section 4.5.6). A drying time of at least 12 weeks should be planned for a 15 cm thick wall lining made of wood-chip light earth, without additional forced drying. Other kinds of light earth may dry a little faster.

Light earth changes colour as it dries and is deemed sufficiently dry to be plastered when the colour no longer changes. To assess whether this is the case it is important to also check material from beneath the surface. One can compare this with pieces of the mixture that are without doubt dry, e.g. dried spillage material. More precise results can be obtained by gravimetric assessment (weighing) of a sample taken from the construction before and after oven-drying. If the light earth wall lining is to be clad with panelling, it is imperative to ensure that the mixture has fully dried.

### 7.2.5 Fixing items to the light earth wall lining

Lightweight items can be fixed to the wall lining with long and strong wood screws ($\geq 6 \times 120$ mm) that are held in place by the structure of the aggregate in the light earth. For larger loads, supporting timber subconstructions should be integrated into the wall construction (see above). Injection anchors can also be used. Recessed electrical socket boxes can be fixed by cutting a suitable hole and fixing them in place with gypsum plaster. For greater stability they can be screwed to the rear wall of the socket box to the underlying structure.

## 7.3 Light earth masonry wall linings

Fig. 7.5 : Construction principle: light earth masonry wall linings

### 7.3.1 Description of the internal insulation

Light earth masonry wall linings are erected in typical brick bond as an insulating layer on the internal face of external walls. It is particularly important to ensure that the cavity between the wall lining and the existing wall is entirely filled and free of voids.

### 7.3.2 Constructing masonry wall linings and appropriate materials

The cavity between the external wall must be thick enough that it can be entirely filled with mortar but not so thick that it cannot dry out sufficiently quickly. Vapour permeable and capillary conductive external walls and wall linings are beneficial for the drying process.

The *Lehmbau Regeln* specify a maximum thickness of 3 cm. Here we recommend a maximum thickness of ca. 1 cm. Should the cavity need to be thicker as a result of specific conditions on site, irregularities in the wall should first be evened out with the help of a levelling layer of earth plaster or earth masonry mortar. Any such layer should first be fully dry before beginning with the masonry wall lining.

Because the wall lining should reduce the thermal transmission of the wall, blocks and mortar with a low lambda-value should be chosen. Stable and uniform light earth blocks (Usage Class I) are available on the market with bulk densities ≥ 700 kg/m³ (chapter 8). They contain a high proportion of mineral or organic lightweight aggregates. Light earth masonry mortar also contains lightweight aggregates. The products available on the market have a bulk density of between ca. 1000 - 1200 kg/m³. In combination this results in a bulk density for the wall lining of ca. 800 kg/m³ and a lambda-value of ca. 0.21 W/mK according to DIN 4108. Light earth blocks are available in different formats. 2DF format is commonly used for internal wall linings.

Masonry wall linings must rest on a stable, structurally-sound base. The cavity between the wall lining and the external wall is filled with each course of blockwork. The wall lining should be securely anchored to the external wall to ensure that the fresh masonry does not separate from the mortar layer. Unstable brickwork increases the risk of voids and defects. The blocks can be fixed with flat wall ties or perforated metal strips that are fixed with wood screws (turned upwards and fixed from above) before application of mortar. As with brick wall linings, 5 fixing points per m² are sufficient. Wire wall ties are of limited use as the comparatively soft and slow drying earth mortar does not provide a firm enough grip. Anchors should be fixed to the external wall on the timber members or where sufficiently stable directly into the panel infill. Wall anchors are necessary if fixing wall ties into brick or stone walls.

Internal wall linings with a slenderness ratio of h/d > 15 must be anchored to the external wall for structural reasons. This means that for an 11.5 cm thick wall leaf, anchors are required for all wall linings over 1.70 m high. The kind of fixture must fulfil the structural requirements.

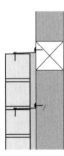

Fig. 7.6 : Anchoring the masonry wall lining to the external wall

Because earth mortars harden only slowly, particularly when the blocks are not very absorbent, there is a limit to how much wall can be erected in one go. The maximum permissible height is 2 m per day. For the same reason and due to the low compressive strength of earth masonry mortar, the mortar joints should be fully filled (chapter 8).

Even the most well executed and detailed masonry wall lining will be subject to settlement. If the walling extends between the timber beams, e.g. the beams pass through it, a sufficient gap should be left between the beam and masonry. If no gap is left the load of the masonry may rest on the timber beam as the wall settles, leading to settlement cracks in the structure. Any gaps that arise at the top of the wall as a result of settlement are to be closed after drying.

The arrangement of installation recesses and chases is described in chapter 9, as is the mounting of recessed electrical wall socket boxes.

## 7.4 Insulation board wall linings

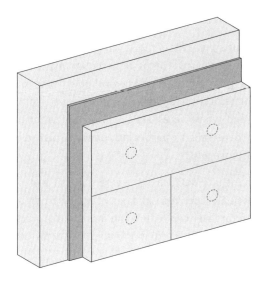

Fig. 7.7 : Construction principle: insulation board wall linings

### 7.4.1 Description of the internal insulation

Insulation boards that provide a suitable plaster base for subsequent earth plaster are applied in a bed of mortar to the internal face of the external wall.

The insulation boards themselves do not consist of earth building materials but the technique is described here because they are bedded in earth mortar. The viscous consistency of earth mortar is ideal for fixing the insulation boards. The way they harden ensures that they form a firm bond even when the underlying surface is absorbent and the mortar is not exposed to air: earth mortars do not suffer the same problems as other mortars in this respect. They are vapour permeable and have a minimal effect on moisture transport between the forward face of the insulation and the existing wall. For this reason, earth mortars are also suitable for use on the internal face adjoining the room, especially where capillary conductive insulation materials are used.

## 7.4.2 Mortar layer, application and fixing of the insulation

For the construction to function properly, it is essential that the layer of mortar between the insulation and external wall is continuous and free of defects. Due to its capillary conduction, it can absorb small quantities of moisture in the wall (caused, for example, by driving rain), drawing it away from timber members. The same applies for interstitial condensation that may arise where warm air meets the face of the insulation. Its main purpose, however, is to ensure a void-free contact between the insulation panel and the external wall. Warm, humid air from the room interior should not be allowed to flow behind the insulation as this would lead to moisture collection through surface condensation within the structure of the wall.

To ensure a continuous transfer of moisture, the layer of mortar must be sufficiently thick. On the other hand, to prevent excessive moisture in the construction, especially where the insulation material contains organic materials, the layer should not be too thick. In the *Lehmbau Regeln* the maximum thickness is given as 3 cm. Here we recommend a maximum thickness of ca. 1 cm and, where necessary, the prior application of a levelling layer of earth plaster mortar where walls are particularly uneven. This levelling layer must be completely dry before proceeding.

Where timber members are exposed on the internal face of the external wall, a plaster lath can be applied to improve adhesion of the wood and earth. The insertion of reinforcement mesh in the adhesive mortar layer can reduce the incidence of crack formation in timber-frame constructions. Fibre-less earth mortars are better for use with reinforcement mesh than fibrous earth mortars.

Fig. 7.8 :
Void-free bedding of insulation board in earth mortar

The insulation boards are applied directly to the viscous earth mortar layer. To ensure good contact between the two, the fresh mortar layer can be combed with a notched trowel. If the boards are to be applied to an already dry levelling layer or other even surface, the insulation boards can be fixed with a thin layer of fine-grain mortar applied to the surface with a notched trowel. The adhesive layer can alternatively be applied to the reverse side of the insulation board. Even better contact is ensured with a combination of the two: a layer of combed adhesive mortar on the wall and the rear face of the insulation board. The application of the panel should be undertaken carefully (figure 7.8).

To ensure full contact between the panel and adhesive mortar layer, most kinds of insulation board are additionally fixed with long screws and washers, impact anchors or insulation fixings of the kind used for external thermal insulation composite systems that firmly press them into the mortar layer. 5 fixing points per m² is generally sufficient. Once fixed in place, they can be plastered over after a short drying time.

### 7.4.3 Kinds of insulation boards

The different kinds of insulation boards used for internal insulation have different technical and physical characteristics. They are summarised in table 7.2. The following widely available kinds of insulation boards can be plastered with earth mortars:

#### Wood-wool lightweight building board

Wood-wool board as defined in DIN 1101 are cement or magnesite-bonded lightweight building boards with an open surface structure. The boards are very vapour permeable but not capillary conductive. They have been applied with mortar as insulation board for decades.

#### Reed insulation board

Reed insulation board consists of a mat of reed stems arranged parallel to the surface of the board. The board width is simultaneously the length of the reed stems. The reed stems are bound together with galvanized wires running perpendicular to the panel surfaces interwoven with wires running the depth of the panel. Reed insulation board is as yet not a certified insulation material but their quality is established with simple methods typically used in agricultural farming. The boards are very vapour permeable. Their capillary conductivity from the front to the back of the panel is very low.

### Wood fibre insulation board

Wood fibre insulation board consists of fine wood fibres that are made into a pulp with water and the formed into boards under heat and pressure. They are a normed building material defined in DIN 68755-1. They are bonded together through the action of the fibres and the lignin released during their manufacture. Some producers use small amounts of additional synthetic resin binding agents. Boards are available in amorphous and multi-ply forms. They are vapour permeable but exhibit only a low degree of capillary conductivity.

### Calcium silicate board

The boards are made of lime slurry, sand and sometimes additional reinforcing fibres. The components are pressed in superheated steam under high pressure. The material has a fine, porous and capillary conductive structure. The surface finish is smooth and absorbent. Due to their alkalinity and good capillary conductivity, they are often used where moisture dissipation and insulation are needed simultaneously. The boards are rigid which means they cannot be pressed to fit slightly uneven subsurfaces.

### Mineral foam insulation board

These boards consist of powdered quartz, hydrated lime and cement which are manufactured using a foaming agent and then steam-cured in autoclaves. The material structure has a limited degree of capillary conductivity and a porous surface, while the bulk density and thermal conductivity are much lower than that of calcium silicate boards. They are not particularly strong and the boards are generally applied to the flat surfaces of massive constructions. The panels cannot be fixed with anchors or similar and the earth mortar must therefore provide sufficient adhesion.

Table 7.2 : Product characteristics and physical properties of selected commercially available types of insulation board

| | Dimensions mm | Thickness mm | Thermal conductivity | Capillary conductivity | Sorption capacity |
|---|---|---|---|---|---|
| Wood-wool building board | 2000 × 500 (600) | 15, 25, 35, 50, 75, 100 | – | – | + |
| Reed mat insulation board | 2000 × 1000 (1250) | 20, 50 | + | – | + |
| Wood fibre insulation board | 1300 × 790 | 30, 40, 60, 80, 100 | + | + | + |

Table 7.2 : continued

|  | Dimensions mm | Thickness mm | Thermal conductivity | Capillary conductivity | Sorption capacity |
|---|---|---|---|---|---|
| Calcium silicate board | 1250 × 1000 [2] | 25, 30, 50 [2] | + | + | + |
| Mineral foam insulation board | 580 × 380 | 60, 80, 100 | + | + | + |

[1] A lower lambda-value is possible at D = 50 mm depending on the quality of the material.
[2] Also available as thin window reveal boards 500 × 250 × 15 mm.

## 7.5 Material and building element properties

### 7.5.1 Thermal insulation and protection against moisture

Table 7.3 : Bulk densities, λ-values and μ-values of selected building materials for internal insulation

| Building material | Bulk density kg/m³ | λ-value W m/K | μ-value |
|---|---|---|---|
| Wood-chip and expanded clay light earth | 600 | 0.17 | 5/10 |
| Wood-chip and expanded clay light earth | 800 | 0.25 | 5/10 |
| Foamed glass light earth | 460 | 0.114* | 5/10 |
| Light earth with kieselguhr | 300 | 0.08* | 5-15* |
| Light earth masonry 700 | 750** | 0.24 | 5/10 |
| Light earth masonry 800 | 830** | 0.27 | 5/10 |
| Light earth masonry 1000 | 1000** | 0.35 | 5/10 |
| Light earth masonry 1200 | 1160** | 0.45 | 5/10 |
| Wood-wool lightweight building board | 360 | 0.090 | 2/5 |
| Reed insulation board | 140-160 | 0.065 | 1-3 |
| Wood fibre insulation board | 180 | 0.045 | 5 |
| Calcium silicate board | 200-240 | 0.065 | 3-6 |
| Mineral foam insulation board | 115-130 | 0.045 | 3-5 |

\* According to manufacturer's details.
\*\* Taking into consideration the proportion of 1000 kg/m³ earth masonry mortar in 11.5 cm thick masonry.

Table 7.4 : U-values with and without typical dimensions of internal insulation. Interstitial condensation calculations may be necessary in individual cases.

| | Existing fabric, no insulation | Light earth 15 cm + earth plaster | Light earth masonry 11.5 + 1 cm + earth plaster | Reed board./Calcium silicate, 5 cm + earth plaster | wood-fibre/ mineral foam 6 cm + earth plaster |
|---|---|---|---|---|---|
| Timber-frame 14 cm, earth panel infill 700 kg/m³, external & internal plaster | 1.20 | 0.58 | 0.68 | 0.60 | 0.45 |
| Timber-frame 14 cm, earth panel infill 1200 kg/m³, external & internal plaster | 1.69 | 0.66 | 0.81 | 0.69 | 0.50 |
| Timber-frame 14 cm, brick panel infill 1600 kg/m³, internal plaster | 1.93 | 0.69 | 0.85 | 0.73 | 0.52 |
| Timber-frame 14 cm, stone infill 2200 kg/m³, internal plaster | 2.66 | 0.74 | 0.94 | 0.77 | 0.54 |
| Massive wall, 24 cm brickwork 1600 kg/m³, internal plaster | 1.82 | 0.69 | 0.85 | 0.73 | 0.52 |
| Massive wall, 36.5 cm brickwork 1600 kg/m³, internal plaster | 1.36 | 0.62 | 0.74 | 0.66 | 0.48 |
| Massive wall, 30 cm stone 1800 kg/m³, internal plaster | 2.82 | 0.80 | 1.02 | 0.86 | 0.58 |

## 7.5.2 Sound insulation

Table 7.5 : Sound reduction index with and without internal insulation

| dB | Existing fabric, no insulation | Light earth 15 cm + earth plaster | Light earth masonry 11.5 + 1 cm + earth plaster | wood-fibre * 6 cm + earth plaster | mineral foam 6 cm + earth plaster |
|---|---|---|---|---|---|
| Timber-frame 14 cm, earth panel infill 700 kg/m³, external & internal plaster | 35 | 43 | 44 | 28 | 38 |
| Timber-frame 14 cm, earth panel infill 1200 kg/m³, external & internal plaster | 41 | 46 | 47 | 34 | 43 |
| Timber-frame 14 cm, brick panel infill 1600 kg/m³, internal plaster | 45 | 49 | 50 | 38 | 47 |
| Timber-frame 14 cm, stone infill 2200 kg/m³, internal plaster | 48 | 51 | 52 | 41 | 49 |

Table 7.5 : continued

| | dB | Existing fabric, no insulation | Light earth 15 cm + earth plaster | Light earth masonry 11.5 + 1 cm + earth plaster | wood-fibre * 6 cm + earth plaster | mineral foam 6 cm + earth plaster |
|---|---|---|---|---|---|---|
| Massive wall, 24 cm brickwork 1600 kg/m³, internal plaster | | 51 | 54 | 54 | 44 | 52 |
| Massive wall, 36.5 cm brickwork 1600 kg/m³, internal plaster | | 56 | 58 | 58 | 49 | 57 |
| Massive wall, 30 cm stone 1800 kg/m³, internal plaster | | 55 | 57 | 57 | 48 | 56 |

(according to approximate calculations conducted by the Schall- und Wärmeinstitut in Alsdorf).
* dynamic stiffness of wood-fibre insulation board, s' 50 MN/m³

## 7.5.3 Fire performance

Table 7.6 : Building material fire safety class of selected building materials for internal insulation

| Building material | Bulk density kg/m³ | Building material fire safety class |
|---|---|---|
| Wood-chip and expanded clay light earth | 600 | B2 |
| Foamed glass light earth | 460 | A |
| Light earth with kieselguhr | 300 | B1 |
| Light earth masonry 700 | 750** | B2 |
| Wood-wool lightweight building board | 360 | B1 |
| Reed insulation board | 140-160 | B2 |
| Wood fibre insulation board | 180 | B2 |
| Calcium silicate board | 200-240 | A |
| Mineral foam insulation board | 115-130 | A |

* According to the Lehmbau Regeln and manufacturer's details.
** Taking into consideration the proportion of 1000 kg/m³ earth masonry mortar in 11.5 cm thick masonry.

# 8 Earth block masonry

## 8.1 Introduction

Earth blocks and bricks are used in new buildings primarily as infill for timber-frame constructions. In Germany, earth blocks are only rarely used for loadbearing masonry constructions.

The following chapter describes earth blocks and earth mortars as building materials for masonry construction in new buildings and renovation work. This chapter focuses on techniques used for new buildings. Their use for insulating wall linings to improve the thermal performance of historical timber-frame and monolithic constructions is discussed in chapter 7. The use of earth block masonry for the replacement and repair of infill panels in timber-frame constructions, as well as in the renovation of monolithic earth structures is discussed in the chapter on the renovation of historical building fabric, chapter 10.

The requirements and properties of earth blocks and earth masonry mortar are given in the German earth building codes, the *Lehmbau Regeln* [DVL, 2008/1]. The Dachverband Lehm e.V. is currently developing a draft norm in conjunction with the Federal Institute for Materials Research and Testing (BAM), which will initially be published in the form of a technical information sheet (TM) entitled "Earth block masonry – definitions, building materials, requirements and testing procedures" (in German: TM Lehmsteine) [DVL, 2011]. The TM describes the quality requirements for earth block masonry in greater detail than in the Lehmbau Regeln. It also elaborates the first clear testing criteria for the classification of earth blocks into Usage Classes. Further technical information sheets for earth mortars are planned. The technical information sheet is scheduled to be published in the first half of 2011 and its content is for the most part agreed. It is therefore incorporated into this chapter. It is not at present possible to say when these will be adopted as a norm.

In Germany, stabilised earth blocks and earth mortars and stabilised masonry constructions are comparatively seldom and there is little experience of corresponding products and their application. As such they are not described in this chapter.

## 8.2 Earth blocks

### 8.2.1 Base material and manufacturing methods

Earth blocks or bricks are made of soils suitable for construction. They can contain mineral (e.g. perlite or brick chippings) and organic (e.g. wood chips) aggregates. Manufacturers should declare any aggregates that are not already a constituent part of the base soil material. Earth blocks can contain synthetic or plant fibres. Here too, the kind of fibres should be declared by the manufacture.

Soluble salts, which can damage the structure of the earth blocks, plaster and coatings as humidity levels change, may only be present in insignificant quantities. This should be ascertained when initially testing the raw material.

Earth blocks can be made in different ways: manually using a hand mould, in a moulding press or rammed or extruded. The manufacturing process has a strong influence on the resulting structure of the block and its properties. The manufacturer should declare the process used.

*Hand-made earth blocks* are made by 'throwing' a suitably malleable earth mixture into a mould and striking off any surplus material by hand or with a wire. The process of filling the mould can also be mechanised.

*Press-moulded earth blocks* are made by compacting a suitable naturally-moist earth mixture in a mould using a stamping press.

*Rammed earth blocks* are made by tamping a suitable naturally-moist earth mixture into a form using a hand-operated or mechanically-driven tamper.

*Extruded earth blocks* are made by forcing a suitably malleable earth mixture through a die and slicing the resulting extrusion into individual bricks.

### 8.2.2 Requirements of earth blocks

#### 8.2.2.1 Usage Classes

Earth blocks are classified according to application area in Usage Classes. Table 8.1 shows the classification into Usage Classes as given in the technical information sheet (TM Lehmsteine) published by the DVL and scheduled for the first half of 2011 [DVL, 2011]. The Usage Classes detailed in the Lehmbau Regeln are identical but lack the additional subclassification of Usage Class I.

Table 8.1: Usage classes of earth blocks as given in the TM Lehmsteine [DVL, 2011]

| Application area | Usage Class |
|---|---|
| Rendered masonry infill of a timber-frame external wall construction exposed to the weather | I a |
| Fully rendered external masonry construction exposed to the weather | I b |
| Clad, weather-protected external masonry construction and interior masonry constructions | II |
| Dry earth block applications (e.g. as ceiling fill or stacked wall lining) | III |

The Lehmbau Regeln [DVL, 2008/1] describes the requirements for earth blocks for each of the Usage Classes as follows:

Earth blocks classified as **Usage Class Ia** and **Ib** should have a homogenous structure, be sufficiently free of residual moisture and frost resistant and should exhibit minimal swell characteristics. If sufficiently strong, they can be used for loadbearing masonry. With the exception of handling recesses, blocks in Usage Class I should not be perforated. The proportion of recesses should not exceed 7.5 %.

Earth blocks classified as **Usage Class II** must be sufficiently homogenous, exhibit a solid structure and should not swell excessively when exposed to moisture during bricklaying and plastering. If sufficiently strong, they can be used for loadbearing masonry. The proportion of perforations in blocks of Usage Class II should not exceed 15 %.

Earth blocks classified as **Usage Class III** should be sufficiently stable for the intended application. There are no limitations on the degree of perforation.

Earth blocks that have no particular Usage Class declaration may only be used for Usage Class III applications.

In the new TM Lehmsteine technical information sheet [DVL, 2011] the requirements for earth blocks have been described in detail for each of the Usage Classes. For the first time, testing criteria are now available which manufacturers can use to classify their products. Up to now, this was largely a matter of experience. The essential aspects of the new, more detailed requirements as well as the testing criteria are detailed below.

### 8.2.2.2 Inner and outer geometry

Earth blocks should take the form of a rectangular volume. Deviations from the rectangular, parallelepiped shape are permissible for hand-formed earth bricks and rammed earth blocks. In such cases, this should be declared by the manufacturer.

The dimensions of earth blocks follow the typical octametric system of units (12.5 cm) and block formats used in Germany (table 8.2).

Table 8.2 : Designation of earth block formats according to [DVL, 2011]

| Format | Length | Breadth | Height | [mm] |
|---|---|---|---|---|
| 1 DF (thin format) | 240 | 115 | 52 | |
| NF (normal format) | 240 | 115 | 71 | |
| 2 DF | 240 | 115 | 113 | |
| 3 DF | 240 | 175 | 113 | |
| 4 DF | 240 | 240 | 113 | |
| 5 DF | 240 | 300 | 113 | |
| 6 DF | 240 | 365 | 113 | |
| 8 DF | 240 | 240 | 238 | |
| 10 DF | 240 | 300 | 238 | |

The nominal size, minimum and maximum dimensions of earth blocks are given in table 8.3. Special formats that deviate from these must be designated accordingly by the manufacturer.

Within a batch delivered to a site, the dimensions of the smallest and largest blocks should not exceed the dimensional tolerance $t$ given in table 8.3. Larger tolerances resulting from the manufacturing process must be declared by the manufacturer.

Perforations in earth blocks of Usage Class I and II must run perpendicular to the horizontal coursing. Perforations in earth blocks classified as Usage Class III may run in any direction – including parallel to the horizontal – as long as they are sufficiently strong for the envisaged application and corresponding installation process.

The proportion of perforations in earth blocks may not exceed the levels given in table 8.4 depending on the respective Usage Class. Handling recesses or grip holes should be considered as perforations though not the mortar pockets. Perforations should be distributed evenly across the horizontal surface; their cross-section can take any form. Perforations (with the exception of handling recesses or grip holes) should not exceed 6 cm². Grip indentations and finger holes are only permissible for 3 DF formats and larger. The cross section of an individual recess or grip hole should not exceed 15 cm². Recesses or grip holes should be arranged in the centre of the brick depth and only where needed for handling. The distance between two recesses or grip holes should be at least 70 mm and not contain any further perforations.

Table 8.3 : Dimensional tolerances of earth blocks according to [DVL, 2011]

| Dimension * | Nominal size mm | Minimum mm | Maximum mm | Tolerance $t$ mm |
|---|---|---|---|---|
| Block length $l$ or Block breadth $b$ | 90 | 85 | 95 | 6 |
| | 115 | 110 | 120 | 7 |
| | 145 | 139 | 148 | 8 |
| | 175 | 168 | 178 | 9 |
| | 240 | 230 | 245 | 12 |
| | 300 | 290 | 308 | 14 |
| | 365 | 355 | 373 | 14 |
| Block height $h$ | 52 | 50 | 54 | 4 |
| | 71 | 68 | 74 | 5 |
| | 113 | 108 | 118 | 6 |
| | 155 | 150 | 160 | 6 |
| | 175 | 170 | 180 | 6 |
| | 238 | 233 | 243 | 7 |

*Earth blocks may also be made with a breadth of 60, 80,100,150, 200, 225, 250 and 275 mm and a length of 190, 210 and 290 mm. The dimensional limits of these factory sizes should be observed analogue to the maximum and minimum dimensions given above.*

Earth blocks with an overall perforation surface of less than or equal to 15 % are termed *solid earth blocks*. Earth blocks with a perforation proportion in excess of 15 % are *perforated* or *hollow earth blocks*.

Depending on the Usage Class, earth blocks should exhibit a minimum web thickness as given in table 8.4.

Table 8.4 : Permissible perforation and minimum web thickness of earth blocks as given in [DVL, 2011]

| Usage Class | Permissible perforation proportion % | Faceshell thickness mm | Web thickness mm |
|---|---|---|---|
| I a | Not perforated * | Not perforated * | Not perforated * |
| I b | 7.5 | 30 | 20 |
| II | 15 | 15 | 4 |
| III | No requirement | No requirement | No requirement |

*with the exception of two centrally arranged grips or handling holes for formats ≥ 3 DF.*

Fig. 8.1: Non-permissible use of earth blocks with a perforation proportion > 15 % for loadbearing masonry

### 8.2.2.3 Bulk density and bulk density classes

Earth blocks are classified according to bulk density classes. When determining the block bulk density, any perforations are neglected. The block bulk densities must lie within the limits defined for the respective bulk density classes in table 8.7.

Earth blocks that fall within the bulk density class 1.2 or lower due to the presence of lightweight aggregates, pores or perforations can be termed light earth blocks.

### 8.2.2.4 Compressive strength and deformation of earth blocks under load

Earth blocks are classified into compressive strength classes (see table 8.8). Earth blocks for use as loadbearing masonry must conform to at least compressive strength class 2. Extruded earth blocks with a high clay content can exhibit a compressive strength of up to 12 N/mm². They are, however, very sensitive to water and therefore classified only as Usage Class III. Usage Class III blocks may not be used for loadbearing construction and the high compressive strength cannot be exploited.

Earth blocks with no declaration of compressive strength class by the manufacturer are automatically classified as compressive strength class 0.

For compressive strength testing, earth blocks with a nominal height of ≤ 71 mm are sawn in half and laid as masonry. For all other earth blocks, the compressive strength is tested with whole blocks. For the mortar joints in the test specimen and to ensure an even compression area (bearing surface) cement mortar with 1 part by volume cement (no particular strength class) and 1 part washed natural sand (0–1 mm) should be used. Alternatively, gypsum can also be used. The contact surface of the specimen

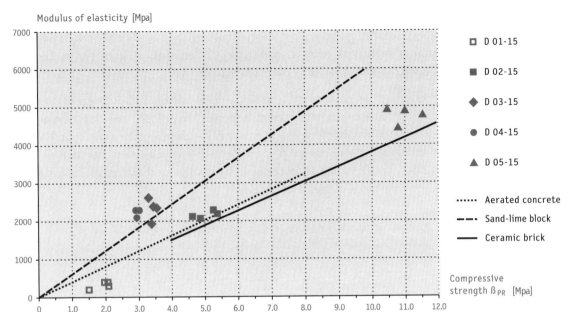

Fig. 8.2 : Stress-strain behaviour of different earth bricks (D 01 to D 05) compared with ceramic bricks, aerated concrete blocks and sand-lime blocks [Haase, 2009]

does not necessarily need a top bed if the bearing surfaces of the test specimen lie plane-parallel and are ground flush. The prepared test specimen is then conditioned at 23°/50 % RH until it reaches a constant weight. The test specimen is then loaded until failure. The rate of loading should be chosen so that failure occurs within 30 to 90 seconds. A test series of at least 6 test specimens should be undertaken.

Earth blocks for loadbearing construction must have a modulus of elasticity of ≥ 750 N/mm². Earth blocks of compressive strength class 2 generally fulfil this criteria. The elastic moduli of commercially available loadbearing earth blocks typically lie between 2000 and 3000 N/mm². As such, earth blocks (for their respective compressive strength) exhibit a similar stress-strain characteristic to other artificial blocks such as fired bricks or aerated concrete bricks (Fig. 8.2). The poisson's ratio of longitudinal to transverse elongation lies in the region of 0.25 to 0.45. The elastic modulus is ascertained using a compression test machine and a suitable measuring apparatus for determining the longitudinal extension of the earth blocks. This is measured in the third load cycle under a load equivalent to a third of the breaking load.

### 8.2.2.5 Moisture and frost resistance

Earth blocks are not moisture or frostproof. Nevertheless, depending on their application, they still need to be resistant to moisture and frost.

Earth blocks classified as Usage Class I may only exhibit a low swelling behaviour. This must also be fulfilled when wet plaster is applied in layers up to 3 cm thick.

Earth blocks classified as Usage Class II should not swell excessively when bricklaying and plastering (layer thicknesses of up to 1.5 cm). Thicker layers of plaster can only be applied when explicitly denoted by the manufacturer.

Testing procedures for moisture and frost behaviour for classification in Usage Classes are not currently described in the Lehmbau Regeln [DVL, 2008/1] and classification has been the responsibility of the manufacturer and a matter of their experience. The new TM Lehmsteine details for the first time a testing procedure and standard values [DVL, 2011]. Table 8.5 summarises the desired performance characteristics of earth blocks of different usage classes under testing.

Table 8.5 : Performance characteristics for testing the moisture and frost behaviour of earth blocks

| Usage Class | Dip test | Contact test | Suction test | Frost test |
|---|---|---|---|---|
| | loss of mass, % | | h | cycles |
| I a | ≤ 5 | No surface cracking or swelling deformation | ≥ 24 | ≥ 5 |
| I b | ≤ 5 | No surface cracking or swelling deformation | ≥ 3 | ≥ 2 |
| II | ≤ 15 | No surface cracking or swelling deformation | ≥ 0.5 | No requirements |
| III | No requirements | No requirements | No requirements | No requirements |

Fig. 8.3a and b : Dip tests with earth blocks of different usage classes (source: BAM)

Fig. 8.4 a and b : Earth blocks after exposure to 0.5 g/cm² water on its top surface (contact test). The cracking shows that the specimen does not fulfil the requirements. (Source: BAM)

For a first appraisal of the water behaviour of an earth block, it is briefly exposed to extreme exposure to water (dip test, figure 8.3). The earth blocks are lowered via an appropriate holding device into a container with water and left in position for 10 minutes. The residue material left in the water is what is measured. Earth blocks of Usage Class I and II may only exhibit a small mass loss (table 8.5).

The test most useful for determining classification in Usage Class II is the *contact test*, which simulates the effect of the application of mortar or plaster to a single surface. A layer of absorbent fabric is soaked in a certain quantity of water and then placed on the facing surface of a block. The amount of water corresponds to the water content of a 1.5 cm thick layer of earth plaster (0.5 g/cm²). An earth block of Usage Class I or II may not exhibit scaly surface cracks or lasting swelling deformation. The earth block shown in figure 8.4 does not fulfil these requirements. Swelling deformations as well as cracking on the surface exposed to water have formed. Furthermore the exposure to moisture has caused the earth block to crack entirely. This earth block is therefore classified for Usage Class III.

The test most useful for determining classification in Usage Class I a and I b is usually the *suction test*. This testing procedure simulates the presence of a quantity of water for a certain time at a certain point, which the block takes up through capillary action (figure 8.5). The simulation corresponds to critical conditions common in timber-frame constructions resulting from moisture ingress or standing water at the boundary between the frame and infill (I a) and the ingress of driving rain through plaster or cracks in the plaster (I b). Earth blocks of Usage Class I a and I b should not exhibit any cracking resulting from swelling or plasticisation over the time periods given in table 8.5. Earth blocks of Usage Class II should likewise be able to sustain such capillary exposure to moisture for a short period and should not crack open on the surface or soften too quickly.

The different kinds of earth blocks, made of different base materials and with different manufacturing techniques can exhibit quite different and sometimes even contradictory behaviour when exposed to moisture through these three tests. All three tests are

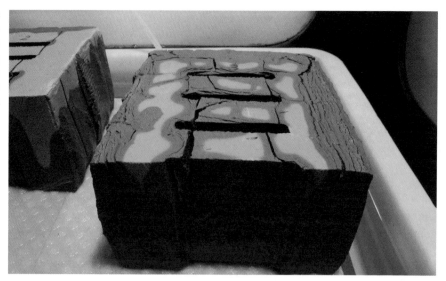

Fig. 8.5 : Suction test conducted on an earth block. The splitting and flaking pattern of failure can be seen clearly (source: BAM).

therefore necessary to be able to ascertain the behaviour and resistance of an earth block under exposure to moisture.

Earth blocks that fulfil the required resistance to moisture for Usage Classes I a or I b are additionally subjected to frost exposure tests. Here the earth blocks are exposed to moisture via an absorbent fabric as described above (figure 8.4) and then placed in a freezer. Earth blocks of Usage Class I a or I b must be able to withstand such exposure for the number of cycles given in table 8.5 without exhibiting surface cracking or lasting swelling deformation.

### 8.2.2.6 Fire performance

Earth blocks must conform at least to the building material class B1 "difficult to ignite" in accordance with DIN 4102-1:1998-05.

Earth blocks without organic aggregates or fibres can be classified as building material class A "non-flammable" without further testing.

In order for an earth block containing organic aggregates or fibres to be classified as building material class A "non-flammable", the product must be tested according to the procedure given in DIN 4102-1:1998-05.

Earth blocks with a material bulk density of $\geq 1000$ kg/m$^3$ can be classified as building material class B1 "difficult to ignite" without further testing. This classification is most probably given due to the fact that hard to ignite mixtures are available with bulk densities of 600 kg/m$^3$ [DVL, 2009].

In order for earth blocks with a material bulk density of < 1000 kg/m³ to be classified as B1 "difficult to ignite", they must be subjected to testing according to the procedure given in DIN 4102-1:1998-05.

## 8.3 Earth masonry mortar

Earth masonry mortar is made of a base earth material and appropriate aggregates and possible fibres. The need for fibrous reinforcement of masonry mortars is not as pronounced as it is for earth plaster mortars as they are subject to less bending and impact stresses and the danger of crack formation is less critical. Earth mortars with a bulk density of ≤ 1200 kg/m³ can be termed light earth mortars.

Earth masonry mortars and earth plaster mortars have many similarities. Earth masonry mortars are therefore described here only in as far as they differ from earth plaster mortars, which are described in detail in chapter 4.

The measure of shrinkage of earth masonry mortars must be reduced to such a degree that no cracks form in the mortar joints. The *Lehmbau Regeln* do not, however, specify a target value for the shrinkage measure that should not be exceeded. For typical mortar joint thicknesses of approx. 1 cm, a shrinkage measure of ≤ 2 to 3 % ensures crack-free joints.

The maximum grain size of earth masonry mortars is typically 2 mm.

The bulk density and shrinkage measure of factory-made earth masonry mortars must be tested in the same way as for earth plaster mortars (see section 4.8). Mortars prepared on site can be assessed using samples and are not required to be tested.

For earth masonry mortars for non-loadbearing masonry and loadbearing masonry up to a permissible compressive stress of 0.3 N/mm², the Lehmbau Regeln do not specify basic minimum requirements for the compressive strength of mortars [DVL, 2009]. By way of orientation, a typical value for mortar compressive strength is 1.5 N/mm². For loadbearing masonry with a permissible compressive stress of > 0.3 N/mm², the mortar compressive strength should approximate the compressive strength of the earth block (table 8.7). This requirement for the compressive strength of masonry mortar can be regarded critically as it lies above the otherwise typical ratio of brick to mortar compressive strengths for masonry.

If lime mortar is used in place of earth mortar for earth block masonry, the same requirements apply as above. Where lime mortar is used, the earth blocks should be pre-wetted to prevent the mortar from drying out too rapidly.

## 8.4 Non-loadbearing earth block masonry with and without timber studs

Non-loadbearing walls and masonry infill are mostly subject to load from their self-weight and, in the case of external walls, also wind loads. Such walls span a single storey. Their stiffening contribution cannot generally be brought to bear on calculations. Non-loadbearing walls must fulfil the requirements given in DIN 4103-1:1984-07 (non-loadbearing partitioning walls). The herein specified fixing (cantilever) loads are generally accommodated without problems as long as the walls are sufficiently thick and resistant to buckling and earth blocks with a bulk density in excess of 1200 kg/m³ are used. Earth blocks classified as Usage Class I or II should be used and laid in earth or lime mortar.

Non-loadbearing earth block walls should have a slenderness ration of h/d ≥ 15 to prevent buckling. Up to a storey height of max. 3.25 m, the ability of the blocks to hold one another is effective up to a distance of 3 m away from the point of anchorage of the wall. That means that walls up to a length of 6 m can be constructed without additional stiffening elements (when anchored at both ends). For longer walls or higher storey heights, an additional stiffening element must be realised to prevent buckling. This is usually achieved with a vertical timber stud. A panel between two studs may not exceed 3 m in length.

Due to the necessary slenderness ratio and to distribute loads, non-loadbearing earth block walls are almost always realised in the form of timber stud walls with storey-height infill panels. For stud walls with the thickness of half a brick, 6/12 cm timber studs are used at 85 cm centres. This leaves sufficient space in the 79 cm wide panels for slightly thicker joints at the edges, which are required for inserting triangular battens.

In freestanding walls, the studs should be positioned, as is usual, either side of openings. The panel infill over the opening can be undertaken , without the need for structural proofs, for openings up to a width of 1.25 m when supported on a 12/12 cm timber lintel. In masonry construction, lintels should have a bearing depth of at least 17.5 cm.

In the case of slender earth brick walls in particular, there is a limit to the height that can be erected each day. The earth mortar must first give off part of its moisture to the earth blocks so that it is sufficiently firm to support the self-weight of the masonry. The maximum height per day is therefore dependent on the suction characteristics of the respective earth blocks used. Depending on the wall thickness, typical values lie somewhere between 1.5 and 2.5 m.

Chases and recesses should not exceed the dimensions given in DIN 1053-1 (design and construction of masonry), Table 10.

Fig. 8.6 :
Non-loadbearing partitioning walls with earth block infill in a private house in Berlin

In many cases the desire to create walls with a high weight per unit area, in order to reduce sound transmission and improve thermal retention, conflicts with structural concerns. As a result, earth blocks with a bulk density of 1200 to 1600 kg/m³ are typically used. The bulk density of the earth block and mortar should be approximately similar.

The sound insulation of non-loadbearing earth block walls without a timber frame should be calculated on the basis of the weight per unit area. Timber stud walls with earth block infill require separate calculation. Partitioning walls with half-brick thick earth block infill generally fulfil the recommendations for normal sound insulation for partitioning walls in residential buildings. Better sound insulation can only be fulfilled by a 17.5 cm thick wall without studs or the use of an additional flexible wall lining (for example earth building boards).

All of the above constructions would, on assessment, be accorded F 30 fire performance. Unfortunately, fire safety certificates for respective building elements and individual constituents are lacking. A similar construction can be found in the model constructions detailed in DIN 4102-4 with a 10/10 cm timber cross section and at least one 1.5 cm thick layer of plaster on one side. This has been classified as F 30 B (30 minutes fire resistance, made of combustible material).

Lightweight fixings can be fixed directly with long screws to fibre-reinforced earth blocks with a bulk density of approx. 1200 to 1400 kg/m². For vertically perforated earth blocks, cavity fixings of the kind used for perforated fired bricks can be used. Medium loads can be fixed in solid earth blocks using normal expanding anchors. For heavier loads injection anchors should be used. The stability of the overall construction should always be taken into account. If in doubt, fixings should be made directly into the timber studs or other scantlings incorporated especially as a subconstruction into the wall construction.

## 8.5 Loadbearing earth block masonry

### 8.5.1 General aspects

Walls and pillars are deemed loadbearing when they support vertical and/or horizontal loads and/or serve as a stiffening construction for loadbearing walls.

In the design of buildings with loadbearing walls made of earth building materials it is necessary to take into account the building material properties and manufacturing technique.

The dimensioning of loadbearing earth walls is described in the Lehmbau Regeln. Their validity is limited to buildings with no more than 2 full storeys and not more than 2 dwelling units. All other cases require special consent on an individual basis from the planning/building control authorities.

At present, the dimensioning of loadbearing earth block walls follows the basic principle of a global safety factor that by lowering the masonry compressive strength is reflected in the permissible stresses. A switch to dimensioning according to partial safety factors in accordance with DIN 1055-100:2007-09 is planned.

The construction of structures with loadbearing earth walls should only be undertaken under the instruction and supervision of a professional experienced in loadbearing earth building construction.

For loadbearing earth block masonry, the general rules and requirements for bricklaying apply, for example adherence to brick bonds and fully filled horizontal and vertical mortar joints.

Earth block masonry walls should not be constructed when there is a danger of frost. During the construction process and up until begin of the warranty period, walls should be protected against moisture ingress from the top and side (e.g. plaster, cladding, roof covering). A commonly overlooked source of water ingress is splash water from scaffolding boards.

## 8.5.2 Construction principles

As the strength of earth building materials is dependent on moisture content, it is essential to prevent moisture and damp penetration of all kinds, not just the surfaces, for the lifetime of the building. Earth block external walls must be protected from rising damp through the incorporation of a horizontal damp proof course above the supporting plinth. The plinth should be made of a material that is unaffected by water or frost. A further, at least 5 cm thick layer of water-insoluble material should be arranged above the damp proof course before beginning with the construction of the actual earth block masonry. This layer is generally made with regular bricks and provides additional protection from water that in certain circumstances, such as damages or poor maintenance, may run down the wall and collect on top of the damp proof course. Earth block internal walls should also be protected against moisture damage from standing water. The water-insoluble material should extend not just 5 cm above the floor slab [DVL, 2008/1] but at least 5 cm above the finished floor level, as it is here that water spillages occur (e.g. from a leaky washing machine).

The mixing of earth block masonry with other masonry building materials in the direction of load dissipation (generally vertical) should be avoided due to the different stiffnesses of the materials which can lead to load redistribution and crack formation. The introduction of horizontal ring beams is, by comparison, unproblematic. A proviso, however, is that these do not cause appreciable stresses as a result of their thermal expansion and contraction. For this the ring beam should be located on the warm side of the insulating layer so that temperature variations are less pronounced.

Chases and recesses in loadbearing earth walls are permissible without further proofs as long as their position and dimensions conform to the stipulations given in DIN 1053-1:1996-11 (Masonry Construction), Table 10. Any recesses in excess of these dimensions must be detailed and taken into consideration in the proof of structural stability.

## 8.5.3 Loadbearing structure and dimensioning

As with any kind of building, loadbearing earth buildings must transfer all horizontal, vertical, permanent and non-permanent loads safely and without non-permissible deformations into the foundation and subgrade. For this there needs to be a sufficient number of loadbearing and stiffening walls, ceilings and ring beams. It is not necessary to provide proof of spatial stability if the arrangement and dimensions of the stiffening walls of the building – in both longitudinal and transverse directions – conform to the details given in table 8.6.

Table 8.6 : Wall thicknesses and maximum intervals between stiffening walls [DVL, 2008/1]

| Thickness of the wall to stiffen cm | Storey height m | Minimum thickness of stiffening crosswall cm | Maximum distance between centres m |
|---|---|---|---|
| 24 to 36.5 | ≤ 3.25 | 11.5 | 4.5 |
| > 36.5 to 49 | ≤ 3.25 | 17.5 | 6.0 |
| > 49 to 61.5 | ≤ 3.50 | 24 | 7.0 |

If the stiffening walls are interrupted by openings, the distance between the opening and the wall to be stiffened should be at least ≥ ¼ to the storey height and not less than 75 cm. The stiffening walls should extend downwards all the way to the foundations or cellar walls without any significant shifts in alignment or other weakening construction that may compromise its stiffening function. To avoid the problem of differential settlement, the stiffening walls should be erected at the same time as the wall to be stiffened. Should their parallel construction be exceptionally difficult, a junction should be pre-formed for the later structural connection of the stiffening wall, for example with an open interlocking brick bond or the embedding of a wall anchor connection rail.

The minimum wall thickness of loadbearing external walls for a maximum permissible storey height of 3.25 m must be at least 36.5 cm. Loadbearing interior walls should also be this thick. When the following conditions all apply, loadbearing internal walls can also be constructed as 24 cm thick masonry:

▸ Storey height ≤ 2.75 m
▸ Live loads, including any supplements for partitioning walls ≤ 2.75 kN/m²
▸ Intermediate supports for continuous ceiling slabs have a bearing distance of ≤ 4 m, ≤ 6 m when employing a centering strip on the ring beam.

Table 8.7 is required for calculating the proof of compressive stresses.

The building materials for the bearing surfaces of ceilings, beams and door and window lintels should be selected to be able to sustain the occurring loads. The distribution of stresses under point loads can be estimated with 60°. Loads should be placed as centrally as possible or else the influence of eccentricity needs to be taken into account in the structural stress analysis.

Ceilings are generally borne by a ring beam. Ring beams should be used where ceilings do not provide a sheet action or structural ring function of their own, which is usually the case for the lightweight timber joist or laminated beam ceilings commonly used in conjunction with earth construction. Ring beams can be made of timber. Because the coefficient of friction between wood and earth is comparatively small but shear forces need to be transferred via the contact surfaces, timber ring beams must be anchored securely with anchor rods deep into the masonry construction. As a result, ring beams are typically made of reinforced concrete, for example embedded in brick lintel blocks (figure 8.1).

The bearing surfaces for lintels should be at least 24 cm deep. Where calculations require longer bearing surfaces the lintel deflection should be limited to l/500. Alternatively the section of wall subject to higher loads can be constructed out of a stronger material, for example brick. Again, a load distribution angle of 60° can be used. For dimensioning lintels, the standard triangular load principle used in masonry construction can be applied.

### 8.5.4 Physical performance of loadbearing earth block walls

The sound insulation of massive earth block walls should be calculated using the weight per unit area. The recommendations for normal and improved sound insulation for the walls of a dwelling are fulfilled with all typical bulk densities. Party walls between dwellings can be constructed of 24 cm thick earth block masonry with a bulk density ≥ 1800 kg/m³ and plastered on both sides.

Earth block walls made of earth blocks with building material class A "non-flammable" with a thickness of 24 cm or greater have a fire performance of F 90 A (90 minutes, non-combustible).

## 8.6 Material and building element properties

Table 8.7 : Block bulk density, bulk density classes and the thermal conductivity coefficient of earth blocks [DVL, 2008/1], [DVL, 2011]

| Bulk density class | Mean value of block bulk density kg/dm³ | Thermal conductivity coefficient, $\lambda_R$    W/mK |
|---|---|---|
| 0.6 | 0.50 to 0.60 * | 0.17 |
| 0.7 | 0.61 to 0.70 * | 0.21 |
| 0.8 | 0.61 to 0.80 * | 0.25 |
| 0.9 | 0.81 to 0.90 * | 0.30 |

| Bulk density class | Mean value of block bulk density kg/dm³ | Thermal conductivity coefficient, $\lambda_R$ W/mK |
|---|---|---|
| 1.0 | 0.91 to 1.00 * | 0.35 |
| 1.2 | 1.01 to 1.20 ** | 0.47 |
| 1.4 | 1.21 to 1.40 ** | 0.59 |
| 1.6 | 1.41 to 1.60 ** | 0.73 |
| 1.8 | 1.61 to 1.80 ** | 0.91 |
| 2.0 | 1.81 to 2.00 ** | 1.10 |
| 2.2 | 2.01 to 2.20 ** | 1.40 |

* Individual values may only exceed or fall below the class limits by not more than 0.05 kg/dm³.
** Individual values may only exceed or fall below the class limits by not more than 0.10 kg/dm³.

Table 8.8 : Compressive strength classes of earth blocks as given in [DVL, 2011]

| Compressive strength class | Compressive strength, N/mm² Mean | Smallest individual value |
|---|---|---|
| 0 | No requirement * | No requirement * |
| 2 | 2.5 | 2.0 |
| 3 | 3.8 | 3.0 |
| 4 | 5.0 | 4.0 |
| 6 | 7.5 | 6.0 |

* Earth blocks of compressive strength class 0 must be sufficiently firm for handling and the intended application.

Table 8.9 : Permissible compressive strength for earth block masonry with earth or lime mortar

| Block compressive strength class | Permissible compressive stress * N/mm² |
|---|---|
| 2 | 0.3 |
| 3 | 0.4 ** |
| 4 | 0.5 ** |

* For pillar-like walls, the permissible stress up to 1.5 times the minimum cross section should be reduced by a factor of 0.8.
** Permissible on proof that the compressive strength of the mortar equates to that of the respective earth block used. Without proof, the permissible stress is 0.3 N/mm².

# 9 Rammed earth construction

## 9.1 Introduction

Rammed earth is used to construct loadbearing and non-loadbearing walls as well as rammed earth floors.

The following chapter describes rammed earth as a building material for walls and floors as well as current construction techniques for making rammed earth. The *Lehmbau Regeln* [DVL, 2008/1] define the requirements and properties of rammed earth as a building material as well as the building elements constructed out of rammed earth. There are very few examples of stabilised rammed earth products or uses in Germany. For this reason the use of stabilisers with rammed earth is not discussed here.

Due in part to a number of high-profile building projects, rammed earth has received comparatively widespread publicity in the press. It is, however, not as widely used in practice, a fact that can be attributed to the high cost of rammed earth constructions.

Fig. 9.1 Chapel of Reconciliation, Berlin

Rammed earth is a massive and monolithic form of earth construction. The resulting constructions are formed directly out of the raw material. Today, rammed earth is used, for example, for south-facing walls or in passive solar energy concepts in conjunction with extensive glazing. Entire buildings have even been constructed to exploit the moisture and thermal retention properties of rammed earth. However, a more likely reason for the renaissance of this old building technique lies in the strong aesthetic and architectonic expression of the heavy and monolithic rammed earth walls. Their presence is a factor of their sheer mass. Rammed earth has been formed into sculptural walls that define space or solid walls that contrast with lightweight and transparent constructions.

## 9.2 Rammed earth

### 9.2.1 Raw materials and manufacture

In the past, rammed earth construction was most widespread where earth of a suitable consistency and granularity was readily available. Today, rammed earth is typically mixed from different aggregate materials specifically for the intended use.

Soils suitable for use as rammed earth should be at least lean in terms of its binding characteristics. Soils with very low natural cohesion should be enriched through the addition of richer earth or powdered clay. Particularly well-suited soils include mountain soils and slope wash which are sufficiently cohesive with a mixed granular to stony consistency. Such soils can sometimes be used directly for rammed earth without further modification. In general, however, the building material is prepared as a mixture. Semi-rich earth mixtures are commonly used because they are sufficiently adhesive to bind together the necessary aggregates to form an ideal grain skeleton. Aggregates include angular particles with rough, split surfaces. The largest grain size should be appropriate for the intended use as well as the desired surface finish. An optimal grain size distribution lies in the region of grading curve B as given in DIN 1045-2: 2008-08.

Organic additives, such as a flax fibres can be added to increase the strength and to counteract crack formation in the material.

The raw material must be carefully mixed to a homogenous mass with a paddle mixer. Tumble mixers are not appropriate.

Mixing is most effective when the raw material is relatively dry (but not dusty). If the material is mixed when it is too plastic, the particles are not sufficiently rubbed together, instead sliding over one another rather than interlocking and spreading. Water is added only once the mixture is properly and evenly mixed, and only as necessary to reach the desired installation moisture content. This can be determined sufficiently precisely by hand (see section 9.3).

Only in exceptional cases are rammed earth mixtures manufactured on site. This chapter therefore concentrates predominantly on the requirements of factory-produced mixtures and their processing.

## 9.2.2 Properties

### 9.2.2.1 Bulk density

The bulk density of rammed earth in a compacted state lies between 1700 and 2400 kg/m³. Rammed earth mixtures with a lightweight aggregate can also exhibit lower bulk densities. The density is tested on a cube with an edge length of 20 cm. The bulk density of the dry building material as supplied by a manufacturer is given in kg/m³ rounded off to the nearest 100 kg/m³. Tolerances of ± 10 % are permissible.

### 9.2.2.2 Measure of shrinkage

The measure of shrinkage of rammed earth should be no more than 2 % and for visible building elements no more than 0.5 to 0.7 %.

The measure of shrinkage is measured using a rammed test specimen with dimensions 600 × 100 × 50 mm (l × b × h) into which two markers are scratched 500 mm apart. A foil is placed inside the mould in which the test specimen is rammed so that adhesion to the walls of the mould cannot affect the degree of shrinkage. After removal of the mould, the test specimen should be allowed to dry unassisted under normal indoor air conditions. The final distance between the two markers is then measured to determine the measure of shrinkage. Before starting construction works, the measure of shrinkage should be checked for the first batch, thereafter every 10 m³ for rammed earth mixed on site and every 50 m³ for factory-mixed rammed earth material. Manufacturers are required to state the measure of shrinkage in % rounded off to one decimal place. The actual measure of shrinkage may not exceed the value stated.

### 9.2.2.3 Compressive strength and modulus of elasticity

Rammed earth for loadbearing applications should have a compressive strength of at least 2 N/mm². For non-loadbearing walls, there is no minimum requirement for compressive strength. The need for sufficient abrasion resistance is given above a compressive strength of approx. 1.5 to 2 N/mm². As a result a compressive strength of 2 N/mm² can be regarded as a threshold value for suitable material. The compressive strength of typical rammed earth mixtures generally lies between 2 and 4 N/mm².

The compressive strength is measured using a minimum of three test cubes. The smallest individual value is taken as the relevant value. The test cubes with an edge length of 20 cm must be made with the same density as the material will be used on site. Forced drying of the test cubes is not permitted. The cubes are tested according to the *Lehmbau Regeln* [DVL, 2008/1] after they have reached their equilibrium moisture level under normal indoor air conditions, usually after around 6 weeks. In future it is expected that the test conditions will be given more explicitly as 23 °C/50 % RH. Before starting construction works, the measure of shrinkage should be checked for the first batch, thereafter every 10 m³ for rammed earth mixed on site and every 50 m³ for factory-mixed rammed earth material. The *Lehmbau Regeln* do not specify a minimum requirement for the modulus of elasticity. Rammed earth for loadbearing applications should, like loadbearing masonry, exhibit a modulus of elasticity of at least 750 N/mm². Mixtures with a compressive strength of 2 N/mm² generally fulfil this requirement.

### 9.2.2.4 Moisture and frost

Rammed earth is not moisture or frost proof and the resistance of rammed earth to the effects of moisture or frost varies can vary considerably. It depends on many factors, such as the kind and quantity of clay minerals in the mixture, the kind of aggregates and the whether or not fibres have been added. In general, richer, fibrous mixtures are more resistant than leaner fibre-less mixtures.

There are as yet no universally agreed testing procedures for assessing the behaviour of rammed earth under exposure to moisture and frost have. It would be desirable to define a series of Usage Classes for different uses of rammed earth, much like those already in use for earth blocks, so that rammed earth mixtures can be classified according to particular test criteria.

### 9.2.2.5 Fire performance

The building material class of rammed earth does not need to be determined according to the *Lehmbau Regeln* [DVL, 2008/1], as the high density of mixtures used for rammed earth are sufficient to be classified as building material class A "non-flammable", even when the mixture contains organic fibres.

## 9.3 Constructing rammed earth walls

### 9.3.1 Introduction

Rammed earth walls should be planned and dimensioned according to the *Lehmbau Regeln* [DVL, 2008/1]. It is imperative to be aware of and to observe the stipulations contained therein. The planning and execution of rammed earth requires a high degree of technical expertise and practical experience. Rammed earth works should be undertaken under the direction of a specialist with adequate experience of the theory and practice of this earth construction method.

In the design and planning of a building project with rammed earth, the architect, structural engineer and construction contractor should be involved from the outset. This ensures on the one hand that the constructional possibilities of this building method are exploited and on the other that unrealistic expectations of its performance are corrected.

For constructions where rammed earth is to be exposed to view, it is advisable, as with all facing finishes, to construct trials of surface finishes sufficiently in advance.

### 9.3.2 Building material preparation

Storage and transportation processes can alter the moisture level and homogeneity of carefully produced factory-mixed rammed earth mixtures. The contractor should therefore not assume that the material can be used without any further processing.

Fig. 9.2 : Hand testing the moisture content prior to installation

In most cases, the material needs to be moderately wetted, which can be achieved simply with a spade and even a water spray. The same process helps in restoring the homogeneity of rammed earth mixtures that may have settled a little. Mixtures that have started to separate more considerably need to be mixed anew with a paddle mixer to ensure that they are sufficiently homogeneous in consistency.

The right degree of moisture can be determined sufficiently precisely by hand. The process involves taking a handful of material and pressing it together. The material should not fall apart, nor stick to one's fingers but remain more or less intact without being particularly malleable.

The rammed earth can only be introduced with a certain tightly-defined level of moisture content. The permissible tolerance is comparatively small. If the material is too wet, it displaces laterally when rammed rather than compacting. When it is very wet, it sticks to the ram or tamper. If, on the other hand, it is too dry, it is hard to compact sufficiently and does not stick together adequately.

The resulting prepared mixture should be covered with a tarpaulin to protect it against drying out or precipitation.

### 9.3.3 Shuttering and formwork

The force of the compaction of the material acting on the formwork is often underestimated. Formwork must be dimensioned to withstand a surface load of 60 kN/m². Shuttering systems of the kind used for concrete construction are suitable for use as

Fig. 9.3 : Uneven layering evident where steel tie rods obstructed the compaction process

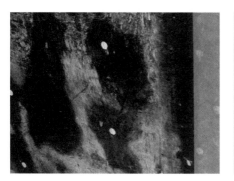 

Fig. 9.4 : Traces of cement from re-used concrete shuttering are visible on the rammed earth surface

formwork, as are own constructions made on site. The shuttering must be anchored securely to the ground to prevent it from "wandering" upwards when compacting. The formwork should be checked at intervals to ensure it remains plumb. If travelling formwork is used, the shuttering elements should be sufficiently large to ensure that a good portion of the existing rammed wall remains shuttered. The degree of overlap depends on the geometry of the element being rammed but should be at least 1 metre. To ensure clean edge details and to reduce the susceptibility of the corners to damage, triangular battens can be laid in the internal corners of the shuttering.

The shuttering should be constructed so that a minimum number of shuttering tie rods are necessary. These get in the way when compacting the mixture, which can result in uneven layering showing on the surface of the wall if adequate care is not taken.

The surface of the shuttering influences the surface of the finished building element. Shuttering boards that have previously been used for concrete should not be used for

Fig. 9.5 : Laminated veneer shuttering boards with film coating inserted inside the basic formwork with offset butt joints

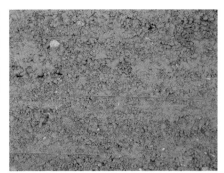 

Fig. 9.6 : Surface of mixture with high aggregate content (left) compared with a typical mixture (right)

rammed earth. Traces of oil or cement that are practically invisible on the shuttering board will show as grey blotches on the surface of the rammed earth (figure 9.4).

The use of thin film-coated laminated veneer shuttering boards on the inner face of the main shuttering has proven effective (figure 9.5) for achieving practically uninterrupted surfaces. The panel joins should be arranged offset from those of the main shuttering.

### 9.3.4 Material insertion and compaction

For the transport and pouring of rammed earth, concrete transporters with a large tailgate can be used. The material should be poured in as evenly as possible and not from a great height to avoid the formation of conical heaps which can lead to segregation of the material. The larger particles roll off the sides of the cone and collect along the edge of the shuttering, where they will later be visible on the surface of the wall. Even when carefully compacted, this results in wall surfaces with a high proportion of coarser aggregate (figure 9.6).

Once the rammed earth mixture has been distributed with a rake, it can be compacted. Manual compaction by hand is an extremely strenuous physical activity and only economically feasible for small elements. The weight of a hand tamper should correspond approximately to the dimensions of its base. Tampers with a base dimension of 12 × 12 cm and a weight of 12 kg have proven effective in terms of weight and handling. For the most part, however, ramming is usually undertaken using hand-held pneumatic rammers. A series of different rammers of different sizes, foot shapes and power ratings are typically needed on site. This makes it possible to compact first with a large foot, then with a smaller foot. Small, easy-to-handle pneumatic rammers with a foot with one chamfered corner to match the triangular batten, can be used to get into corners and complex shapes. Larger compaction machinery such as tamping rollers, more commonly used in road building, exercise greater compaction but also greater load on the shuttering. Their use is only effective for long, uninterrupted stretches of wall. In all other cases the effort required to repeatedly hoist the machine into place and the stronger shuttering required are disproportionately high compared with the speed gain. And even when such machines are used, it is still necessary to compact the edge sections along the shuttering with hand-operated pneumatic rammers in order to ensure a uniform pattern of compaction on the face of the wall.

Each layer should consist of 10 to 15 cm of rammed earth, depending on the compaction device used and grain distribution. The mass is then compacted by about 1/3. Each layer must be compacted uniformly and several times over until no further compaction is possible and the surface produced is flat and almost sealed. For constructions that will be exposed, it is important to remember that every stage of work (shuttering, filling in of material, compaction) will be visible on the subsequent wall surface and therefore needs to be undertaken with care.

It is important that compaction of a layer takes place uniformly in several individual stages. To begin with the entire layer is lightly pre-compacted, thereafter re-compacted solidly and finally post-compacted along the edges. Along the course of the wall, one should begin in the middle and work one's way toward the ends of the wall or a corner. This avoids the corners and ends from seeming to taper off.

If geogrid is to be used as reinforcement and to prevent crack formation (section 9.4.3), the first two thirds of the material should first be inserted, the geogrid then laid horizontally before the remaining third of the material is filled in. Only then should it be compacted.

Fig. 9.7 : Compaction with a hand-held pneumatic rammer

### 9.3.5 Removal of formwork and touching up

Rammed earth walls up to a slenderness ratio of h/d = 10 can be constructed in one operation up to a height of one storey (3.25 m according to the Lehmbau Regeln) and the shuttering then removed. Above a slenderness ratio of around 7 the shuttering should be removed in sections and a "corset" put in its place to prevent buckling of the wall. This is usually achieved using anchored vertical timbers placed at 1.5 metre centres. Such constructions are absolutely necessary for cases where a second storey is to be added where the wall below has not fully dried out. In such cases it is advisable to consult an experienced engineer with appropriate measuring equipment to appraise the structural stability of the construction and the rate at which new load can be added to it. Due to their very long drying times, rammed earth walls are not able to sustain their full design load for a long time. It is important to realise that the construction period itself is a critical phase, and one in which the risks are often underestimated.

Directly after removing the shuttering, any possible holes and defects should be repaired. The entire surface should then be brushed down with a soft brush. If desired for optical reasons, working the surface with a soft wire brush can bring out the texture of the aggregate. If the wall is to be plastered, this also helps to provide a good mechanical key.

### 9.3.6 Drying

The drying of rammed earth walls also entails that the building element will shrink. This can lead to crack formation and deformations if the rammed earth mix exhibits a large measure of shrinkage or the structure is allowed to dry irregularly and too quickly. For example, one-sided exposure to strong sunlight should be avoided. All the same, the drying period should be kept as short as possible.

Thin cracks in still moist material can often be closed simply by applying pressure to the edge of the cracks. Wider cracks are generally closed once the structure has fully dried out and is no longer subject to shrinkage. The surface of the construction, which is often already dry, should be carefully wetted in the region of the crack. Even when very carefully executed, the repair of cracks and defects are still visible on the surface of the wall. As such, priority should always be given to avoiding crack formation through the choice of an appropriate mixture and reinforcement against cracking.

During the building and drying period, rammed earth walls, in particular those with surfaces that will be exposed to view, should be protected against rain.

Rammed earth walls should only be constructed when there is no danger of frost. The frost-free period also applies to the drying time. Only when the core of the wall has

Fig. 9.8 : Touching up shrinkage cracks

dried sufficiently is the danger of frost no longer a problem. It is not possible to specify a precise tolerable moisture content as this varies depending on the type of mixture.

The formation of mould on rammed earth wall surfaces occurs very rarely. The material already has a low moisture content when rammed and the surface dries quickly if the drying conditions are sufficiently good. Nevertheless, it is advisable to monitor the wall surface during drying for possible mould formation.

## 9.4 The design of rammed earth walls

### 9.4.1 Constructional measures for weather protection

While rammed earth surfaces are often surprisingly resistant against driving rain, it is not advisable to leave rammed earth walls exposed to the weather in temperate climates such as that of central and northern Europe. When damp sections of wall are subjected to alternating cycles of frost and thaw, serious erosion to a greater or lesser degree will result (figure 9.9). Only where surfaces very rarely experience driving rain (in many regions the north-east face) can external rammed earth walls be realised with a tolerable degree of risk. This decision should only be taken after careful study of the situation on site, and agreed contractually with the clients.

A durable hydrophobic coating of rammed earth surfaces exposed to the weather is not possible. The frostproof stabilisation of a wall material is only possible with the addition of stabilisers, for example cement. This, however, is very rare in Germany.

Fig. 9.9 : Erosion of a rammed earth wall exposed to moisture from driving rain after several frost-thaw cycles

One means of improving the erosion resistance of rammed earth is through the insertion of bands of brick or tile rammed into the mixture at intervals (section 10.2.2).

External render applied directly to rammed earth walls is virtually unknown in practice. Rammed earth walls are almost always consciously constructed to be exposed to view and can be protected against the weather by sufficiently wide eaves or transparent covering. In some cases rammed earth walls are clad with insulation board on the outside and only visible from the interior. The external render should be appropriate for the insulation material and the low diffusion resistance of the rammed earth. Typically softwood insulation board or reed insulation panels are used with a lime-bonded plaster.

Rammed earth walls should always be placed on a foundation of massive, waterproof material. In the case of external walls, this foundation or plinth should be raised high enough ($\geq 30$ cm) to protect the base of the rammed earth wall against splash water. Rammed earth walls likewise need to be protected against rising damp through the introduction of a damp proof course above the plinth. Directly on top of the damp proof course, there should be a layer of waterproof material at least 5 cm high before the actual rammed earth construction begins. This layer is usually made of concrete screed and prevents the collection of moisture from rainwater on the top surface of the damp proof course. It also protects the damp proof course from being punctured when ramming by sharp aggregate particles in the rammed earth mixture.

## 9.4.2 Embedded elements

The combination of materials of different stiffness in a wall can lead to crack formation under fluctuating loads as well as continuous loads. This phenomenon only occurs when the loads within the materials, or at the transitions between the materials, cause stresses of relevant magnitude. Stresses resulting from dead loads and live loads, as well as those caused by temperature fluctuations can be of relevant magnitude. The combination of loadbearing rammed earth with other massive building materials in the direction of load distribution should therefore be avoided wherever possible. By contrast, the introduction of stiff horizontal elements such as ring anchors and ring beams is less critical as the stresses created by wind loads are comparatively small. Here the risk of cracking due to temperature fluctuations is the greater risk.

Lintels can be made of steel girders, timber beams, precast concrete or reinforced in-situ concrete. If shallow brick lintels are to be used, these must be supported until the rammed earth has fully hardened. These are dimensioned for a load distribution arc in the wall above the lintel, however this can only form when the wall has hardened. The support of lintels also needs careful attention as it requires ongoing adjustment as the wall shrinks.

The bearing surface for the lintel must be sufficiently large. The top of walls and the bearing surface for ceilings are usually stabilised with a ring beam or tie. Ring beams are necessary in all cases where the ceiling itself has no ring or sheet effect, which is the case for the lightweight timber joist ceilings or laminated board ceilings often used in conjunction with rammed earth constructions. Ring beams and anchors must be dimensioned according to structural requirements.

Ring anchors or beams can be made of wood. As the friction coefficient between wood and earth are low but thrust forces need to be transferred via their contact surfaces, wood ring anchors and beams should be anchored sufficiently deep within the rammed earth with anchor rods. This is achieved by ramming in anchor rods that have a sufficiently large head plate at their lower end during the construction process.

In most cases, however, reinforced concrete ring anchors or beams are used. Where the wall is sufficiently thick, they can be embedded within the cross-section of the rammed earth so that they are not visible. The space for the concrete ring beam must formed with an additional piece for formwork, i.e. the walls either side of the ring beam are rammed first before the ring beam is cast. Cement seeping from the concrete into the rammed earth ensures a good connection between the two.

Any necessary inserts or wall anchors must be arranged at a sufficient distance to the edges and ends of walls. Expansion anchors with a small spread should be arranged at least 20 cm away, those with a wide spread at least 50 cm. Holes for anchors within 30 cm of the edge of a rammed earth construction should be drilled without using the hammer function. Expansion anchors should not receive tensile loads, only inclined and transverse tension loads. The recommended load for simple walls anchors is just 0.15 kN (15 kg), for frame anchors 0.25 kN (25 kg). If larger loads or tension forces need to be sustained, injection anchors (embedded metal threads bonded in place) should be used. For walls with a thickness of at least 20 cm, these can sustain up to 0.8 kN (80 kg) [Ziegert, 2006]. If wall anchors cannot be used because the distance between them is too small, trapezoidal sections of hardwood or concrete should be embedded in the wall. In the case of sections of hardwood, these must first be wetted before insertion otherwise the moisture within the earth mixture will make the wood swell and expand, causing cracking in the rammed earth.

### 9.4.3 Reinforcement

Rammed earth constructions can be reinforced using geogrids. Geogrids are inexpensive coarse plastic meshes normally used to stabilise soils in dams and embankments. They are made of polyamide-coated glass fibre. Meshes made of hemp, flax or hessian are not strong enough for such uses.

Fig. 9.10 : Reinforcement for a ring beam embedded in a rammed earth wall construction so that it is concealed from view

Geogrids make it possible to reinforce rammed earth constructions against shrinkage-induced cracking. When arranged regularly, they can also increase the loadbearing capacity of the wall. As yet, however, no reliable, systematic testing has been undertaken and it is therefore not possible to factor in the contribution of geogrids in structural calculations for the dimensions of rammed earth constructions.

Geogrids can be used in all situations where the risk of cracking is to be expected, for example areas where the thickness or height of the cross-section is reduced, such as above and below openings or in the remaining cross-sections of niches, or at corners or junctions with other walls. As a rule of thumb, geogrids should be inserted horizontally at a distance of 20 cm above one another. The strips should be laid alternately 50 or 80 cm over the potential location of the crack.

Geogrids are likewise recommended in the region of bearings for point loads to improve the resistance to transverse tensile loads and reduce the risk of cracking. At least three horizontal layers should be inserted 15 cm beneath one another. The length of the strips should be designed so that the ends extend approximately 30 cm beyond the load distribution angle of 60°.

The mesh size should be approximately twice that of the largest grain particles. The ribs of the mesh should not be too wide to avoid creating a separating layer. Where the ribs have different dimensions, the thicker ribs should be arranged parallel to the potential crack. They help resist the shear component of vertical tensile forces above the crack.

See section 9.3.4 for details of installing geogrids. Geogrids can be cut with a cutter on a hard base.

Fig. 9.11 : A roll of geogitter (left) cut into L-shapes as corner reinforcement (right)

### 9.4.4 Installations

The insertion of horizontal installation channels is best achieved by inserting a sufficiently rigid U-profile into the shuttering. To avoid problems during shrinkage, these should not be too high. Vertical channels can be created by placing slightly tapered vertical timbers in the shuttering. The timbers should not be too dry. Similarly electrical conduits can also be embedded vertically or horizontally provided they are sufficiently robust. Special care is required when embedding horizontal conduits.

### 9.4.5 Internal finishes of rammed earth wall surfaces

If the surface of a rammed earth wall needs additional stabilisation, this can be achieved with diluted painter's water-glass or a casein-based binder. It is strongly recommended to conduct trials before applying the stabiliser.

The internal face of rammed earth walls are rarely plastered. Earth plasters can be applied without the need for any special preparation of the subsurface. Before applying a lime plaster, a better mechanical key can be created by brushing the surface with a wire brush or another roughening device to reveal the coarser aggregate. A roughened earth undercoat plaster as a base coat for a subsequent lime topcoat plaster is not advisable as it forms a soft layer between the two stiffer layers of rammed earth and lime plaster.

Rammed earth walls can be tiled. Careful precautions should be taken to protect against possible moisture ingress.

Fig. 9.12 : Perforated metal strips as wall anchors to restrain a rammed earth wall lining.

## 9.5 Non-loadbearing rammed earth walls

Non-loadbearing rammed earth walls are often designed and realised as freestanding sculptural elements. The requirements are accordingly varied and it is difficult to detail specific rules for their construction. Nevertheless for reasons of construction they should not be any thinner than 20 cm. Similarly they should be sufficiently stable. Special care is required when removing the shuttering from slender wall constructions. Wall linings placed up against other wall constructions can be made slightly thinner. They should be anchored to the adjoining wall with wall ties, perforated metal strip (3 to 5 per m²) or geogrid strips. Here too special care is required when removing the shuttering.

## 9.6 Loadbearing rammed earth walls

In the design of buildings with loadbearing walls made with any earth building materials, it is necessary to take into account the particular properties of the material and its manufacturing process. This is especially important for rammed earth constructions.

The dimensioning of loadbearing earth walls is described in the *Lehmbau Regeln*. This applies for all buildings with no more than 2 storeys and not more than 2 dwelling units. In all other cases special consent must be sought on an individual basis from the planning / building control authorities.

The dimensioning of loadbearing rammed earth walls is undertaken according to the concept of a global safety factor which reflects the impact of reduced compressive strength on the permissible loads. A switch to dimensioning with partial safety factors according to DIN 1055-100:2007-09 *Basis of design, safety concept and design rules* is planned.

Loadbearing external walls must be at least 32.5 cm thick. Loadbearing internal walls can be thinner – at least 24 cm thick – under certain circumstances (see section 8.5.3). The permissible stresses for different compressive strengths are given in table 9.1.

Table 9.1 : Permissible compressive stress of rammed earth building elements

| Compressive strength N/mm² | Permissible compressive stress* N/mm² |
| --- | --- |
| 2 | 0.3 |
| 3 | 0.4 |
| 4 | 0.5 |

*For pillar-like walls, the permissible stress up to 1.5 times the minimum cross-section, should be reduced by a factor of 0.8.

Buildings with loadbearing rammed earth walls may only be constructed under the direction and supervision of a suitably experienced specialist in the erection of rammed earth constructions.

The strength of a construction is not solely dependent on the quality of the rammed earth used, but also on maintaining the optimal moisture content during installation and a consistent degree of compaction. For this reason building control authorities have sometimes prescribed that the compressive strength be monitored on site. Non-destructive testing can be undertaken with a Schmidt-Hammer (type PT), a rebound hammer commonly used to test concrete. To properly assess the rebound values measured, calibration measurements first need to be taken on test cubes of known compressive strength.

## 9.7 Rammed earth floors

Rammed earth floors are often employed today in prestige buildings for their visual qualities. They are expected to be hard-wearing and to resist abrasion to a degree not required of them in the past. Historically, rammed earth floors were mostly used for barn floors and other ancillary rooms.

In the lower layers of rammed earth floors, rammed earth with an arbitrary maximum grain size can be used. For the upper layers, the use of mixtures with a maximum grain size of no more than 8 mm is recommended in order to obtain finishes with low surface porosity without the need for extra finishing.

Well-graded earth mixtures are more stable, especially when subjected to point loads.

If the measure of shrinkage of the rammed earth mixture is greater than 0.5 %, crack formation is to be expected. The surface will then require repeated treatment over the drying process to close the cracks.

Rammed earth does not act as a capillary break. When rammed directly on the ground, it can transport large quantities of moisture and will dissipate this into the interior of the room. This property can be exploited for wine and vegetable cellars. For all other applications, however, a damp proof course of bituminous or polyethylene material should be laid on the subsurface. Rammed earth should not be applied directly on top of the damp proof course to avoid the membrane being punctured by sharp-edged particles. As an intermediary layer, 4 cm of hydraulic lime mortar or cement screed can be used. Note that the use of a layer of gravel or foamed glass aggregate as a capillary break is not on its own sufficient to ensure a dry floor. Even the moist air that collects in the capillary breaking layer can still transport enough moisture from the ground to the rammed earth floor for a cardboard box placed on the rammed earth floor to form mould.

If the rammed earth floor is to be installed on a rigid concrete slab or ceiling, the height of the rammed earth should be at least 10 cm, 6 cm for the lower layer and 4 cm for the upper layer. The thickness of the upper layer should not be thinner than 4 cm to avoid the risk of the top layer flaking off in scales.

If the rammed earth is to be installed where there is no comparable rigid slab as a base, the rammed earth itself has to act as a sheet. For normal loads, a construction depth of 20 cm is necessary, 2 × 8 cm for the lower layers and 4 cm for the upper layer.

While it is not necessary for the lower layer to have fully dried out before applying the next layer, any crack formation in the lower layer must be fully completed. Should the cracks be deeper than half of the layer thickness, the cracks should be filled with earth mortar. If the earth cracks apart into shallow concave segments, the rammed earth mixture is unsuitable and must be removed.

A geogrid can be incorporated into the layer beneath the top layer and serves to reduce cracking over larger surface areas or where walls or corners project into the room, both during construction as well as in general use. As with walls, the first 2/3 of the rammed earth layer should be filled in loosely, then the geogrid, then the final 1/3 of the mixture. The entire layer can then be compacted.

The insertion of edging strips along the perimeter walls also helps to counteract crack formation on the surface of the floor. Cork strips of the kind used when installing parquet flooring have proven suitable.

After compaction, the underlying layer should be brushed clean with a hard brush to remove any loose material and in preparation for the next layer.

The pipework of underfloor heating systems should not be embedded directly in rammed earth. Instead they should bedded in hydraulic lime mortar or cement mortar. The thickness of rammed earth above this layer should be at least 10 cm. As a heat radiation system, the overall system is effective but has a very slow response time.

To achieve as flat a surface as possible, the loose rammed earth material can be poured over a series of slender battens that act as a guide for raking level the mixture. The battens are then removed. As with rammed earth walls, compaction takes place in several operations, each across the whole area of the floor. The lower layer can be compacted using hand-held pneumatic rammers or small-scale road rolling equipment. Suitably flat surfaces for the top layer are best undertaken with hand tampers. Absolutely flat floors are generally not possible without an excessive amount of effort.

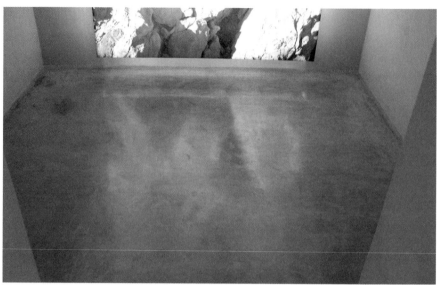

Fig. 9.13 : Waxed rammed earth floor, Jahili Fort, Al Ain, UAE

After compaction, small-scale finishing work such as smoothing uneven surfaces can be undertaken by applying substantial force with a trowel. Lightly spraying the surface with water makes working the material easier but also changes the surface structure, causing it to become less firm. Such work, if at all necessary, should be undertaken sparingly.

By waxing the surface with hard floor wax, rammed earth floors are given a protective hard-wearing surface. Waxed surfaces also look more refined. Oiling the surface with a suitable floor oil prior to waxing further hardens the uppermost surface of the floor. This is recommended for surfaces subject to high wear and tear but also darkens the surface considerably.

## 9.8 Physical properties of rammed earth

Due to their high density and thickness – a product of their manufacture and technical structure – walls made of rammed earth have a high thermal retention capacity but also high thermal conductivity. The commonly implemented single leaf construction with exposed rammed earth surfaces on both inner and outer face is not able to fulfil the minimum thermal insulation requirements given in DIN 4108 Part 2 at any realistic thickness. To fulfil the stipulations of the German Energy Conservation Regulations (EnEV), additional insulation materials are always necessary (table 9.3). External insulation is preferable to internal insulation or core insulation. The realisation of core insulation is a complex and involved process.

With a water vapour diffusion resistance factor of 5/10, rammed earth is an exceptionally breathable building material. If the surface is to be clad for insulation or weather protection reasons, the material used must be chosen with the properties of rammed earth in mind. Insulation materials, plasters and coatings with a high water vapour diffusion resistance factor should not be used on the outer face of the wall. For homogenous rammed earth constructions, the formation of condensation is not a problem.

Rammed earth mixtures are typically wholly mineral in composition. On occasions they contain small quantities of organic plant fibres. All commercially available rammed earth mixtures fulfil the conditions of building material class A. This should be declared by the manufacture.

According to the *Lehmbau Regeln*, a rammed earth wall with a density of at least 1700 kg/m³ and a wall thickness of at least 24 cm can be classified as fire safety class F90. According to currently available data this level of fire resistance is already given for a small wall thickness, and is much higher for a 24 cm thick wall.

Due to their high density – a product of their manufacture and technical structure – walls made of rammed earth have good sound insulation properties. The weighted sound reduction index of the respective construction can be calculated, as with all massive constructions, according to DIN 4109 Supplement 1, Table 1 (table 9.4 in this chapter) using the mass per unit area. Due to the material's low modulus of elasticity, the values obtained in practice are often better than the values given in the table.

## 9.9 Building element properties

Table 9.2 : Coefficient of thermal conductivity of rammed earth

| Bulk density kg/m³ | Coefficient of thermal conductivity W/mK |
|---|---|
| 1700 | 0.82 |
| 1800 | 0.91 |
| 1900 | 1.00 |
| 2000 | 1.10 |
| 2100 | 1.20 |
| 2200 | 1.40 |
| 2300 | 1.50 |
| 2400 | 1.60 |

Table 9.3 : Construction of rammed earth walls with a thermal conductivity of 0.28 W/m²K (minimum requirement for external walls according to German Energy Conservation Regulations, EnEV 2009)

| Rammed earth (density 2200 kg/m³) cm | Insulation (0.04 W/mK) cm |
|---|---|
| 32.5 | 14 |
| 40 | 13 |
| 60 | 12 |
| 90 | 11 |
| 480* | 0 |

*for information only

Table 9.4 : Weighted sound reduction index of rammed earth walls

| Rammed earth (density 2200 kg/m³) cm | $R'_w$ dB |
|---|---|
| 24 | 56 |
| 32.5 | 57 |
| 60 | ≥ 57 |

# 10 Renovation

## – historical earth buildings

## 10.1 Introduction

Until the second half of the 19th century, wood, stone and earth were among the most commonly used building materials. The building methods used were dictated largely by the building materials that were available locally or could be obtained with reasonable effort. Many different regions have been shaped by the distinctive vernacular building traditions employed.

Earth was employed for domestic buildings in many different ways: for making mortars, plasters and floors, as infill in timber-frame constructions or between ceiling beams and even as roof covering. Some of these applications can still be found in a large number of buildings today. Less well-known is that earth was also used until well into the second half of the 19th century as a monolithic building material for load-bearing walls and that monolithic earth construction was the dominant form of earth construction in many rural areas. There are three principle methods: *weller earth construction*, *rammed earth construction* and *earth brick construction*. The onset of industrialisation and the economic production of bricks on a large scale led to a steady decline in the use of massive earth construction methods. Only in the years immediately following the first and second world wars did earth construction experience a renaissance which lasted up to 15 years.

The increase in general interest in earth as a building material has also brought about a greater awareness of surviving historical earth constructions, not only as a legacy of traditional architecture, but also for the opportunity they present, when well-maintained, for conversion with comparatively simple means into places to live and work in made of sustainable building materials.

In the renovation and conversion of historical building constructions, it is essential to possess good knowledge of the materials typically used, their properties and construction methods as well as the typical defects mechanisms they exhibit. These serve as a fundamental basis for evaluating and where necessary modifying buildings in their structure, in their material constitution and their aesthetic appearance. In the follow-

Fig. 10.1: Weller house near Mücheln/ Saxony-Anhalt, photo from 2005

ing chapter, we have chosen to concentrate on the construction methods from the centuries-old tradition of earth construction in the region of what is now Germany, that through their continued existence or renovation potential play a role in current earth building practice.

## 10.2 Solid earth wall constructions

### 10.2.1 Weller earth construction

Compared with modern construction technology, weller earth construction is an almost archaic building method. But it is precisely because of its simplicity – a spade and pitchfork are the only tools one needs – that it has been used over hundreds of years and is the most widespread historical monolithic construction method used in Germany altogether. The oldest known surviving weller buildings date back to around the middle of the 17th century and are most widespread in Central Germany in what are now the federal states of Saxony, Saxony-Anhalt and Thuringia. The technique was most popular in the first half of the 19th century. In the post-war years after the first and second world wars, weller represented only a very small proportion of the overall extent of earth building.

Weller constructions are often confused with other methods, both technically as well as linguistically. Half-timbered constructions with earth infill are sometimes mistakenly called weller-walls, perhaps because of the straw-clay material used, which in Low German was known as "Weller" [Grimm/Grimm, 1885]. More commonly weller constructions are mistakenly identified as rammed earth for the simple reason that weller construction is comparatively unknown today. The shuttering boards used in historical rammed earth construction were typically between 50 and 80 cm high. The layers of weller construction are of a similar height and their seams are therefore often mistaken for shuttering marks. A further reason for the misinterpretation is the presence of holes in the wall that were left over after the removal of wooden scaffolding. Similar holes can also be found in rammed earth constructions: some of these also once held scaffolding, but the majority are left over from the horizontal members that held together both sides of the shuttering. In rammed earth these can be found above every horizontal joint; in weller construction holes were only needed to hold the scaffolding. A clear means of differentiation is the presence of 'ledges' of brick tiles or mortar strips along the surface of the wall which were placed in the formwork of rammed earth constructions to provide a key for later plaster application (figures 10.24 and 10.25). The surface of weller constructions exhibit sliced off pieces of straw, often turned downwards by the slicing action (figure 10.5).

Fig. 10.2 :
Application of the second layer of weller mass, Meti School, Bangladesh 2005

Fig. 10.3 : Traces of the process of weller construction on the surface of a barn in Albersroda / Saxony-Anhalt, photo: 1998

Weller earth walls are made of a stiff mass of straw-clay applied vigorously by hand – i. e. without formwork – using a pitchfork in layers of between 70 and 80 cm high. The vigorous application of the malleable mass provides compaction while the straw content of the mixture prevents it from slumping and running off. The person applying the mass stands on the previous section of wall (or the foundation or plinth when applying the first layer) and applies the mass diagonally working backwards (figure 10.2).

In most cases the weller mass is piled diagonally in layers, changing direction with each layer. This can often be seen on the surface of many surviving weller buildings (figure 10.3).

During application some of the material protrudes over both side of the wall. This can be trimmed off straight – after about 1 - 3 weeks, depending on the weather conditions – using a special straight-edged spade (figure 10.4). In some regions the side faces of each layer are additionally compacted before or after trimming by beating them with a board. The trimmed material can be mixed back into the mass for the next layer. Working without formwork means that it is easy to form rounded corners and curved walls or other geometric forms that are difficult to create with formwork (figures 10.6 and 10.8). Each layer is completed around the entire perimeter of the building before beginning with the next layer on top of the trimmed and hardened wall at the point where the previous layer was started. The earth builder was entrusted with deciding whether the completed layer was solid enough to take the weight of the next layer or whether to wait. In practice, the mass of the subsequent layer was generally less of a problem than the point load of the workman standing on the wall to apply the mass.

Fig. 10.4 : Trimming the surface of the first layer of weller, Meti School, Bangladesh 2005

Fig. 10.5 : Surface of the trimmed weller showing sliced straw stems, Meti School, Bangladesh 2005

Due to its simplicity, weller earth construction was a most rational construction method. The first proper systematic measurement of working times for different earth constructions undertaken after the second world war [Bauwelt, 1949] showed clearly that rammed earth or earth brick construction took about 25 % longer than equivalent weller constructions. Rammed earth requires the construction of shuttering as well as compaction of the mass; earth brick construction requires bricklaying in brick bond and the cutting to size of individual bricks. With the advent of shuttering systems and mechanisation of the ramming process, these figures shifted in favour of rammed earth. At that time, earth brick construction played only a secondary role.

For the few new weller building projects, mechanisation has really only been able to save labour in the preparation, mixing and transport of material. The most labour-intensive aspect, the application of the material and the trimming of the walls, still has to be done manually.

A detailed description of historical weller earth construction is available in [Ziegert, 2003].

### 10.2.1.1 Weller earth and its properties

Weller earth construction is to be found in all regions where suitable and easy-to-work loess soils are present in large quantities. Such soils generally have a very lean to lean consistency, in rare cases semi-rich. Very rich soils and clay as well as very stony soils are unsuitable for weller construction. In some cases, the soil excavated for the foundation and cellar can be used directly for construction. Mineralogical analyses of earth samples from weller constructions along with associated samples taken from

Fig. 10.6 (left): Window rabbet sliced out of the weller mass

Fig. 10.7 (right): Weller mass with twigs as reinforcement in Steinstücken near Berlin, photo: 2005

the site showed in all cases an almost identical composition of the mixture [Ziegert, 2003]. This result can be taken as evidence that, aside from plant fibres, no other mineral additives are mixed into soils for weller construction. In some cases the presence of irregular fragments of brick, gravel or slag pressed into the surface with the intention of providing a better mechanical key for subsequent plastering have been falsely interpreted as a high proportion of stony aggregates.

If further soil was required for construction, this was obtained from public pits. Many village ponds are in fact former clay and earth pits. In earlier times, it was regarded as a fundamental right for villagers to extract earth from the village pit [Simons, 1965].

Left to weather over winter in a moist state, the soil was worked into a stiff, pasty consistency by adding water and treading. "Straw should then be strewn over the earth mass and trodden-in in such a way that the straw fibres fall individually and not in bunches. The more care taken to evenly distribute the straw, the stronger the resulting walls" [Geinitz, 1822]. The preparation of the earth as well as the treading in of straw was a matter for the whole family, servants and even neighbours. Where available, cows and horses were also pressed into service [Miller/Grigutsch, 1947].

Depending on the kind of soil used, somewhere between 22 and 28 kg/m³ of straw should be mixed in. The more cohesive the soil is, the more straw needs to be mixed in. The dry building material has a bulk density of 1400 to 1700 kg/m³. Measurements undertaken on numerous buildings have found bulk densities mostly in the region of 1450 to 1550 kg/m³. The bulk density of weller earth is therefore a little higher than that of the straw-clay used for panel infill in timber-frame constructions (section 10.3.1.1). For the most part rye straw was used, which is relatively sturdy, but wheat straw ranging from 30 to 60 cm in length was also used. "Builders who build solid weller walls chop the straw once, those that want to make their life easier, chop it twice. The shorter straw, however, is no use for this kind of building." [Lange, 1779]. In some regions, fine brushwood, twigs or heather is mixed in instead of straw (figure 10.7).

Samples taken from historical weller earth buildings exhibited compressive strengths of between 0.6 and 1.3 N/mm². With a modulus of elasticity of approx. 300 N/mm², weller earth is one of the softest materials for monolithic construction altogether [Ziegert, 2003]. The strength of mixtures in existing buildings can be measured on site using a resiliometer (rebound hammer SUSPA PT). Calibration measurements have showed that using this type of meter, readings of 35, 40 and 45 correspond to compressive strengths of 0.7 N/mm², 1.0 N/mm² and 1.3 N/mm² [Ziegert, 2003].

### 10.2.1.2 The construction of weller earth buildings

The use of weller earth construction was restricted primarily to the perimeter walls of buildings due to the technique's necessarily thick wall construction. Only in rare cases were internal walls also built using the weller technique. Loadbearing and non-loadbearing internal walls of weller constructions were mostly constructed out of earth bricks. The thicknesses varied depending on the brick or block format used but were rarely more than 30 cm, even for loadbearing walls. The earth brick walls were probably first erected after the shrinkage of the weller walls was more or less complete. The earth brick walls are therefore not structurally connected with the weller walls. In most cases they were inserted into a notch-like recess, rarely more than 5 cm deep, cut into the internal face of the weller wall, but sometimes they were simply placed up against the wall. In either case, this simple junction permitted the transfer of compression stresses from the wall to be stiffened to the stiffening wall. If the external wall was subject to wind suction, however, the cross-walls could not serve a stiffening function due to the lack of a structural bond.

In many of the more common weller barn constructions, stiffening walls are rarely found. The external walls were stiffened with pier buttresses. These were very often located at points where the horizontal load of the roof was transferred into the walls (figure 10.8). Such buttresses were constructed in the same operation as the wall. The length and thickness of the projection typically corresponded to the thickness of the wall to be stiffened. According to current regulations, the design length of the stiffening wall is a factor of the thickness and actual length of the wall to be stiffened. The total length must be at least 1/5 of the storey height. While the buttresses in surviving weller barn constructions do not always fulfil this, they are never significantly shorter.

In the majority of weller construction, the construction methods is only used for the ground floor; the upper storeys are continued as a timber-frame construction (figure 10.9). The origins of this constructional division may lie, among other reasons,

Fig. 10.8 (left): Pier buttress in a weller barn in Saxony-Anhalt, photo: 1999

Fig. 10.9 (right): Barn and cowshed with timber-frame construction resting on a weller ground floor in Heuersdorf, photo: 2007

in the ordinances of the 16th and 17th centuries which decreed that to reduce the amount of timber consumption at least the ground floors should be made as massive constructions (figures 10.10 and 10.11) [ThürLandOrd, 1556], [Fiedler, 1965]. Only in the core areas of weller construction around Leipzig and Halle does one find numerous buildings where the upper storey is also constructed using the weller technique (figures 10.1 and 10.10). These two-storey weller buildings date back to the beginning of the 19th century.

The foundation depth of almost all historical weller buildings is by today's standards very often insufficient. Nevertheless, settlement damage resulting from this has been documented only in a few isolated cases. In the majority of cases both the depth of the foundation as well as the height of the plinth exhibits little relation to the building's function, height or wall thickness. Instead these seem more dependent on the wealth of the original owner and the age of the building. In buildings erected by citizens of lower standing, the foundation is often only 20-30 cm deep, whereas buildings built by more prosperous farmers or in the grounds of estates, exhibit foundation depths of more than 80 cm. The depth of the foundation also increases the later the buildings were built. The same applies for the height of the plinth. Depending on what was regionally available, the foundations and plinths were made of fieldstone, pieces of hard rock or dressed blocks of sandstone or limestone. Masonry mortar made of lime mortar was only found in a few buildings from the second half of the 19th century. In most other cases earth mortar was used and the joints of the plinth then covered with lime mortar. It seems that in rural areas lime mortars only began to replace earth mortars towards the end of the 19th century. In many of the buildings erected after 1850, the entire plinth, or at least the uppermost layer of the plinth was built out of brick masonry. In some weller buildings from the 20th century, the foundation and plinth are made of lean concrete.

Vertical damp proofing as commonly used today were not found in any of the weller buildings examined. An effective horizontal damp proof course made of bituminous paper was found in some weller buildings built in the 20th century. In several detailed examinations of buildings from the 18th and 19th centuries, compacted layers of clay were found beneath the foundation. The effectiveness of such layers was limited at best and in any case moisture was still able to penetrate the foundation from the sides.

The typical thickness of weller walls is around 50 to 60 cm, although many of the buildings curiously had walls that were exactly 52 cm thick. The wall thickness is the same around the entire perimeter regardless of the direction of span of the ceiling and therefore the load distribution. In the few instances where the internal walls were also created as weller walls, the thickness was typically between 40 and 55 cm. The slenderness ratio of the weller walls is never greater than l = 10. In the existing buildings, there does not seem to be a clear relationship between wall height and thickness but rather a progressive reduction of the wall thickness over time. The wall thickness of weller buildings from the early 18th century were between 80 and 110 cm on the ground floor; by the end of the 19th century, this had reduced to 60 or even 40 cm. This was the case regardless of whether the wall above continued as a weller or a timber-frame construction. Likewise there seems to be no relationship between wall thickness and economic prosperity of the owner (see above).

The individual layers of weller are between 50 and 110 cm tall, but usually between 70 and 80 cm. In the existing buildings examined, the ratio of weller layer height to wall thickness lay between 1.3:1 and 1.8:1. New experience in the building of weller constructions has shown that a ratio of 1.5:1 should not be exceeded.

Earth weller walls are generally built with vertical external and internal surfaces. In a small region north of Leipzig there are weller buildings, some of them two-storey, with walls that taper markedly (figure 10.12). The walls grow thinner on the outer surface by more than 5 cm per metre wall height. The internal surface, however, is vertical. The walls of weller barns are always vertical. The tapering walls make sense from a structural point of the view, but are problematic in terms of weathering. Most of these buildings have therefore been given a thick layer of external render.

The long drying time required for weller constructions and the concomitant shrinkage of the weller mass was often seen as a disadvantage. Aside from the poor room climate in the interior during the drying period, it extended the time to completion considerably. Weller buildings should only be rendered after one year, ideally two years. On the one hand plastering too early prolongs the already long time required for the wall to dry. On the other hand, because wall shrinkage and creep is still taking place, the

Fig. 10.10 (left): Tapering weller building in Gottschaina / Saxony, photo: 1998

Fig. 10.11 (right): Cross-section showing the tapered weller construction

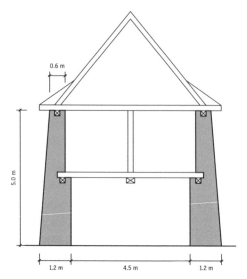

render starts to break and flake off [Fauth, 1933]. Amazingly, despite the long drying time within the thickness of the wall, the straw enclosed within the mixture has not disintegrated. It seems that the structure of the building materials was sufficiently air-free to prevent the straw from humifying. In tests, walls of up to 1.2 m thick were opened up to the centre of the wall. With the exception of a slight coloration, the straw found within was mostly golden-yellow and resistant to tearing.

To improve the adhesion of external render, a mechanical key for the plaster was created by pressing in fragments of brick or slag (figures 10.12 a and b). Other buildings have been textured with patterns of grooves or stipples.

It is not clear to what degree weller buildings were plastered at all in the past. The surfaces of weller constructions are generally more resistant to erosion through wind and rain as a result of their location and the quantity of straw they contain. More probably, however, the situation varies from region to region, for which there are numerous indications in the available literature. *Düttmann* describe villages in the Dübener Heide (between Leipzig and Wittenberg in Central Germany) as being very picturesque due to the warm tone of the many unrendered weller buildings [Düttmann, 1920]. *Bosslet* recounts that in the district of Rehau in Bavaria, weller houses were very rarely rendered, and when, only the newer buildings were rendered or limewashed. The reason, according to *Bosslet*, was lack of money [Bosslet, 1920]. *Meinert* mentions the creation of large roof overhangs for unrendered earth buildings, but recommends coating them with a daub of "... mortar or hair mortar ...", a mixture of lime, clay and hair [Meinert, 1802]. By comparison *Geinitz* recounts: "In this region, the earth walls are coated on the outside with a covering of lime, earth and several other ingredients,

Fig. 10.12 a und b: After trimming the weller wall, pieces of broken brick or slag are pressed into the wet mass to provide a key for plastering

creating not only an impression of a stone wall but also protecting against the vagaries of the weather." [Geinitz, 1822]. This last quotation probably most clearly reflects why owners who could afford to, chose to render their earth walls: it had less to do with protection against the weather and more to do with lending their weller buildings a more cultivated appearance. For this reason, dwellings were more likely to be plastered than those that served simpler functions. This is, at least, the picture we see today. There are comparatively few houses that haven't been rendered and very few farm buildings that have. With a few exceptions, most boundary walls that face roads have been plastered while numerous boundary walls that adjoin fields have not. Both the literature as well as findings on location show that the lime render used for weller buildings was a very thin layer, in principle a little more than a whitewash. *Wrede* recommends the following render build-up that has been used for years in Thuringia: an earth undercoat plaster with a little lime and chaff or flax fibre with a layer of broom-applied lime plaster on top [Wrede, 1920]. *Broom application* involved dipping a besom broom into a slurry and the knocking it against a piece of wood just in front of the wall so that the slurry sprayed onto the wall. It took some practice to achieve an even covering. Samples taken of an external render surface from around that time from a house and stable building in Deumen in Saxony-Anhalt confirm the aforementioned structure of a reinforced earth undercoat plaster mixed with chopped straw and lime coated with a thin lime facing render. For barns, the actual plaster layer – which was responsible for the smooth outer surface – was omitted and the lime slurry applied directly to the weller wall. A large proportion of the renders found on weller buildings today are very thick layers of lime-cement renders. Such plasters often form a skin that with time totally separates from the plaster base. On the internal wall surface, by contrast, the original plaster build-up is very often intact. These are predominantly thin lime finishing plasters applied over an earth undercoat. Sometimes they are also earth finishing plasters reinforced with flax fibres. Barns were not plastered inside. The internal faces of buildings for livestock were given a coat of lime for hygienic reasons: "in cowsheds and sheepfolds, the animals harm themselves by licking the earth walls when they are not coated with lime" [Geinitz, 1822].

An interesting aspect of weller wall construction is the way in which openings were formed. Small openings up to about 1 m wide were often cut out of the dry wall after construction: "Space for doors and windows were not left out but were rather knocked through after the entire perimeter walls had been completed." [Wähler, 1940]. The windows and door openings were, however, generally formed out of earth block, and later brick, arched lintels. Wooden lintels are comparatively rare. Accordingly door and window rabbets were rarely made of wood, but sometimes carved directly out of the weller wall. Occasionally window and door surrounds made of stone can be found (fig-

Fig. 10.13 a to c: Various kinds of lintels over openings: brick, wooden staves and natural stone

ures 10.13 a to c). In many cases the reveals are splayed, becoming wider on the inside to allow more light into the interior.

On most window and door openings it is possible to trace the following process: the transom of the intended opening lies at the same height as the top surface of the penultimate layer of weller. The space for the door is left out in approximate dimensions. After the weller has hardened sufficiently, the window reveal and a notch for the rabbet made of brick masonry, stone or wood are cut out of the weller wall. After construction of the window rabbet and the lintel, the final layer of weller can be applied to the wall. In other buildings the apex of the masonry lintel is at the same level as the height of the wall. In this case, the wall is built to its full height leaving space for the opening. Only then are the window reveals, rabbet and lintel cut and finished. It is also conceivable that the wall plate, ceiling, upper storey and roof were first completed before undertaking the window reveals and lintel. That would provide more time for the earth to shrink before the comparatively rigid masonry rebates were constructed.

The openings in barns for the barn doors are spanned by lintels made of wood, very occasionally of brick masonry. In some barns the section of wall above the entrance was realised as a timber-frame construction to reduce the load resting on the lintel. The comparatively high concentrated loads acting on the bearing points of the lintel were distributed by varying means. In the simplest variant, the beam lies directly on the weller earth construction which has not led to cracks or deformations in any of the buildings examined. In other buildings, cross-pieces made of wood were introduced to spread the load. But in most of the examples researched, the load was borne by timber posts or pier-like constructions made of brick that also doubled as the door reveal. The introduction of stiffer building materials in the direction of load distribution has in some cases brought about shear cracks as a result of shrinkage and creep (figure 10.14).

Individual loads from ceiling joists and roof beams are transferred to the weller walls via wall plates either inserted into the mixture or rested on top. They are often bedded on pieces of rock, brick and straw-clay that serves to even out the inconsistences of the weller layer below. In some buildings, larger pieces of stone are placed beneath the wall plate at precisely the points where individual loads are transferred into the wall (figure 10.9).

Where the upper storey of a weller building was built as a timber-frame construction, the wall plate lies in the outer third of the wall but never flush with the outside surface. This position is obviously a compromise between the structurally ideal location in the centre of the wall and the more weather-resistant location adjoining the external surface of the wall. The wall plates are connected to one another with halved and tabled joints and can therefore – at least to a certain extent – serve the function of a ring beam (figure 10.15).

Fig. 10.14 (left): Shearing cracks over an opening flanked by stiff brickwork piers

Fig. 10.15 (right): Transition from a weller ground floor to a timber-frame upper storey in a weller building from 1658 in the former village of Großgrimma, 1999

In buildings with two full storeys made as a weller construction, the wall plate at first floor level lies primarily on the internal side of the wall but, again, never flush with the surface. In some of these buildings the ceiling joists are simply laid directly on the weller wall without any wall plate.

The wall plates of the upper storey ceiling were laid on the trimmed crown of the wall, or more rarely in a notch cut into the external half of the wall. Horizontal wind loads acting on the roof structure could be transferred to the weller walls via friction between the wall plate and the weller earth. However, this connection was not sufficiently shear-resistant to sustain larger horizontal forces, for example resulting from differential settlement of the ground.

In most weller constructions, the timber joist ceiling is filled with rolled earth reels. In some dwellings with integral livestock housing, the areas for animals have been given a shallow vaulted ceiling.

The chimneys in the weller buildings examined are all without exception built of brick. The brick side walls of the chimney are often bonded with the earth brickwork of the internal walls, but sometimes simply abutted. That the chimneys were not also made of earth may seem perplexing, given that earth has good fire resistance. However, the burning of moist material in the fire creates condensation which increases the moisture content of the earth bricks in turn weakening them. In addition, the problem of blocking and spalling as a result of salt formation is a further problem. This last problem is also suffered by weakly fired bricks to a similar degree.

## 10.2.2 Historical rammed earth construction

The first evidence of rammed earth construction in the region of what is now Germany dates back to the period of Roman settlement (figure 10.16) [Kienzle/Ziegert, 2008]. It was a tradition, however, that was not continued after the end of the Roman Empire.

The second period in which rammed earth regained popularity in what is now Germany was around 1800 as a method imported from France. So enthusiastic was the reception that a number of documents from this period proclaimed the salvation-bringing potential of this new building technique [Seebaß, 1803]. Progressive landowners were the first to employ rammed earth for buildings on their estates but otherwise the method was taken up only reluctantly. The advantages of the new building method were, it seems, not convincing enough to displace traditional regional building methods. It is only through the actions of certain individuals that rammed earth experienced a renaissance in the 19th century in specific localised areas of Germany. A key figure was the industrialist Wilhelm Jacob Wimpf from Weilburg, under whose direction a number of high-quality residential and commercial rammed earth buildings were constructed in the city of Weilburg (figures 10.17 a and b). At that time, these buildings were rammed entirely by hand.

After the second world war, the dearth of industrially-produced building materials led to the revival of rammed earth construction, especially in the Soviet occupation zone, what was later to become the GDR. Through the use of electric and pneumatic rammers it became possible to mechanise the laborious process of compaction and to ra-

Fig. 10.16 : Reconstruction of Roman rammed earth buildings in Xanten Archaeological Park, 2008

 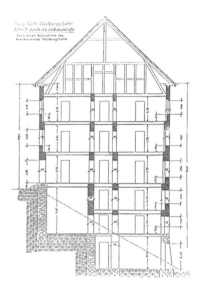

Fig. 10.17 a and b: The tallest massive rammed earth building in Germany, in Weilburg (Lahn), photo and section

tionalise the construction method. As the production of bricks, lime and cement fell far behind demand, rammed earth was used as a means of building urgently needed residential and commercial buildings. In 1947, the Soviet Military Administration in Germany (SMAD) passed decree 209 which ordered the building of 37,000 new farmsteads by the end of 1948 with the express instruction that they be built of "locally available materials" [SMAD, 1947]. According to estimates, some 18,000 monolithic earth buildings – mostly made of rammed earth – were erected in 1948 alone. With the founding of the GDR in 1949, the building of individual rural buildings continued unabated and was expanded to include two-storey buildings in small settlements (figures 10.18 and 10.19) as well as buildings with public functions. But by the mid-1950s, rammed earth construction began to dwindle. Probably the last rammed earth building from this period is a 2-storey rammed earth building built in 1959 in Prädikow in Brandenburg, which was modern and had large window openings for its time (figure 10.19).

In rammed earth construction, a naturally-moist earth mixture is poured in layers of 10 - 15 cm into formwork and compacted. The kind of shuttering used is generally still visible on the surface of the wall. Typically *segmental shuttering* was used with a height of ca. 60 - 80 cm and length of ca. 1.8 m. Larger sections were shuttered in one piece in the buildings erected after the war and the wooden connecting timbers were replaced with threaded steel tie bars.

For single family houses a single layer was shuttered in 4 sections, where each section always turned a corner. The junction between the individual segments was designed as an intentional tension release point. The segments were arranged alternately so that a kind of bond resulted. The shrinkage of the first layer had to have finished before

Fig. 10.18a and b : Complete 2-storey estate in Mücheln/ Saxony-Anhalt

the next layer was applied to avoid the crack in the layer below propagating through the segments of the following layers (figure 10.20b). The corners were particularly susceptible to cracking as were the sections above and below openings. In the reconstruction of the Roman rammed earth buildings in *Xanten Archaelogical Park* battens were laid diagonally within the individual segments to prevent the segments from cracking apart, especially when the construction process proceeded swiftly.

Fig. 10.19 : Probably the last rammed earth building in the former GDR in Prädikow/Brandenburg, built in 1959, photo: 2005

Fig. 10.20 a and b: Historical shuttering and intentional staggered joints where the segments meet

### 10.2.2.1 Rammed earth and its properties

As stony, mountain soil is especially well-suited for rammed earth construction, historical rammed earth is most widespread in regions were such soils are particularly prevalent.

Rammed earth construction is also, however, found in regions where soils contain practically no stony constituents. These overlap with some of the regions in which weller construction is found – Saxony, Saxony-Anhalt, Thuringia as well as some regions on the Lower Rhine – primarily due to the presence of loess soils. Moreover, the stone-free soil is also used for rammed earth without the addition of aggregates. Samples taken from Roman rammed earth in Xanten and an earth pit nearby exhibited an almost identical grain and mineral composition [Kienzle/Ziegert, 2008]. Its use in naturally-moist form along with compaction reduced the measure of shrinkage to within tolerable limits. By using relatively small segmental shuttering, the degree of shrinkage can be accommodated by the regular joints between segments without excessive stresses arising in the material. Only when the shuttering segments grow longer, and the risk of uncontrollable shrinkage cracking grows, is the mixing in of shrinkage-reducing aggregates necessary.

The rammed earth mixture we know today, which consists of a mixture of construction soils and stony aggregates for leaning down the mixture, only became more widespread since the mid-1940s. Because in the post-war period, both suitable materials as well as means of transport were lacking, these mixtures sometimes contain less favourable constituents such as round-grain aggregates instead of angular aggregate material. The addition of straw is not typical.

The bulk density of historic rammed earth mixtures lies between 1600 and 2200 kg/m³ depending on the aggregates contained and the degree of compaction. Rammed loess soils without aggregates have a bulk density of between 1900 and 2100 kg/m³.

The compressive strength of historical rammed earth typically lies between 2 and 3 N/mm², but can sometimes be as low as only 1.5 N/mm². The modulus of elasticity can be estimated at 600 N/mm². The strength of mixtures in existing buildings can be measured on site using a resiliometer (rebound hammer SUSPA PT). Calibration measurements have showed that using this type of meter, readings of 50, 60 and 70 correspond to compressive strengths of 1.5 N/mm², 2.0 N/mm² and 2.5 N/mm².

### 10.2.2.2 The construction of rammed earth buildings

Numerous aspects of rammed earth construction, such as the plinth construction, are similar in principle to those already described in detail for weller construction and are therefore not described again here (see section 10.2.1.2). The following description examines those aspects that differ from weller construction.

Unlike weller earth construction, rammed earth construction is never found in combination with a lightweight timber-frame upper storey. Two-storey rammed earth buildings consist either of a rammed earth upper storey on top of a ground floor made of stone or – more commonly – a ground floor and upper storey both made of rammed earth. The triangular gable ends of buildings are sometimes completed in earth brick masonry or weller construction.

The typical wall thickness of a single or double-storey rammed earth building, and also of the barns built in the 19th century are 50 to 60 cm. Rammed earth walls built after the second world war are often only 40 cm thick.

The height of the individual rammed segments was between approximately 60 and 80 cm, the length between 2 and 4 m.

The rammed earth buildings built in the 1950s all have a horizontal damp proof course in the form of bituminous paper which was protected against damage during ramming by covering it with a course of brick masonry (figure 10.21).

In the level of the ceiling, the sections of wall between the joists were often carried out in earth or ceramic brickwork due to the inordinate effort required to construct shuttering around the ceiling joists.

Openings were generally crowned with a shallow brickwork arch or a timber lintel. In some of the rammed earth buildings constructed in the GDR in the late 1950s, lintels were cast in in-situ concrete. The windows themselves were fixed to the rammed earth via conical hardwood blocks that were inserted into the mixture during the ramming of the wall (figure 10.22). Where twin-leaf box windows were used, the outer set

Fig. 10.21 (left): Proper plinth detail showing horizontal damp proof course, Lindenberg/Brandenburg

Fig. 10.22 (right): Wooden lintel and wedge-shaped blocks inserted into the wall for fixing the windows, Lindenberg/Brandenburg

of windows were often arranged flush with the outside wall to reduce the incidence of splash water at the junction between window sill and reveal. In buildings where the window plane was set back from the face of the wall, this junction was often executed as at least two course of brickwork.

In some rammed earth barns, the section of wall over the barn doors was realised using weller construction instead of rammed earth. By changing the building method, the rest of the still hardening rammed earth structure was not subject to such excessive vibrations. Furthermore, the lintel only needed to bear the smaller weight of the weller wall.

Fig. 10.23: Rammed earth barn with weller earth sections over the entrances in Jüdendorf/Saxony-Anhalt, 2006

Fig. 10.24 a and b: Rows of embedded brick and tile visible on the surface of rammed earth walls

In the majority of historic rammed earth buildings, one can see elements that have been rammed into the mixture to serve as a key for better render adhesion. Typically these were flat pieces of stone or broken roof tiles that were laid in the shuttering on top of a freshly rammed layer before the material for the next layer was filled in (figures 10.24 a and b).

In later buildings, these were instead replaced with strips of lime mortar. For this, lime mortar was inserted along the shuttering and formed into a wedge shape that rose towards the external face (figure 10.25). When no external render was applied, the

Fig. 10.25 a and b: Strips of lime mortar embedded in the surface of rammed earth walls

Fig. 10.26 a and b: L-shaped brick inlays on the surface of a rammed earth wall

pieces of stone or roof tile also helped protect the rammed earth wall against erosion. In this respect, however, the wedge-shaped lime-mortar strips are problematic in that they actually channel rainwater from the wall's surface into the inner structure of the rammed earth wall. It is therefore especially important that the external render is intact on walls where this particular method was used.

A special variant employed in the 1950s was the insertion of L-shaped ceramic profiles instead of tiles into the shuttering, which created the appearance of a badly executed wall (figure 10.26). Irrespective of their appearance, they helped achieve excellent adhesion of the render. In these buildings, the original render is still mostly intact and functional to this day. It is rare to find a building where the render has been damaged, revealing the construction beneath (figure 10.26b).

### 10.2.3 Earth brick construction

Earth brick construction is one of the oldest building techniques known to man and is known to have been used in Germany since the time of Roman settlement [Heimberg/Riesche, 1998]. Finds from other parts of Germany show that timber buildings with earth infill or stone constructions were otherwise more prevalent at that time.

Compared with weller and rammed earth construction, earth brick construction is most similar to our contemporary understanding of building. Of the three monolithic construction methods, however, it is the least widespread. This may have been due to the fact that the small-scale prefabrication of the earth bricks and subsequent assembly into a solid, thick earth wall was more involved than making a monolithic wall

Fig. 10.27 : Historical earth brick wall laid in header bond, Jenaprißnitz, 2004

out of a malleable or naturally-moist earth mass (see the comparison of working times in section 10.2.1). But weller and rammed earth constructions needed a lot of space and could only be built within a relatively narrow time frame during the year. As a result, those building in the more densely built parts of villages or small towns turned to earth brick construction, while buildings at the edges of settlements employed weller or rammed earth construction.

Earth brick construction is also more widespread in regions where the local soils were not suitable for weller or rammed earth construction without laborious leaning-down of the mixture. With earth brick construction, the material shrinkage took place before the wall was built, making it possible to use much richer soils as a raw material.

The historical technique of constructing earth brick walls was no different to that of making conventional ceramic brick walls. Very early surviving examples of earth brick walls (second half of the 17th century) exhibit for the most part a straightforward header bond, while later earth brick walls used block bond or cross bond (figure 10.27).

A special variant of earth brick building is the so-called Dünner earth loaf technique. Gustav von Bodelschwingh, the so-called "earth building pastor" of Dünne in Westfalia, initiated an extensive social project to tackle the shortage of housing. Between 1926 and 1933 almost 300 earth buildings were constructed as self-build projects. The future inhabitants, including the women and children produced earth bricks in the form of loaves and built simple but utilitarian buildings out of them. Most of the buildings still exist today. Bodelschwingh also built a model settlement of houses in park-like surroundings in Bünde-Dünne that is now a listed monument (figures 10.28a and b).

Fig. 10.28 a and b: Unrendered earth-loaf building in Dünne/Westfalen, 1999

### 10.2.3.1 Earth bricks and earth masonry mortar

Earth bricks can be made of practically all kinds of soils without the need for the significant addition of aggregates. Although some soils were subject to considerable shrinkage, this did not present a problem as the bricks were first assembled into a wall when dry, that is after shrinkage had taken place. Where additives were used, these were mostly small proportions of chopped straw or flax fibres.

Up until the middle of the 19th century, earth bricks were made exclusively by hand. A viscous mass of pre-prepared earth was thrown with a vigorous swing into a pre-wetted wooden mould without a base. Where necessary the material was pressed into the corners and then drawn flush at the top. In the German language, hand-formed bricks are known as *"patzen"*, a name possibly derived from the sound of the earth striking the mould. The wooden frame was then drawn upwards leaving the earth brick to dry. The bricks were generally formed on an area of flat ground strewn with sand. The bricks could dry and shrink *without constraint*. After about 3 days the hardened bricks were turned on end, saving space and ensuring the bricks didn't dry out unevenly and misshapen. With the advent of industrial brick production, extruded earth bricks also started appearing in constructions of this time, although hand-made bricks continued to be made and used.

The format of historical hand-made earth bricks varies considerably. In general the size of the bricks reduced over the centuries. The formats range from as large as 400 × 190 × 190 mm (l × b × h) down to the current NF (Normal Format) dimensions (240 × 115 × 71 mm). Extruded earth bricks were usually slightly larger than the *Reichs-*

Fig. 10.29 :
Earth brick masonry wall showing earth masonry mortar reinforced with flax fibres, Domsen 1999

*format* (250 × 120 × 64 mm). The reason for this is that fired bricks were manufactured slightly larger than the final format to compensate for shrinkage during firing.

The density of earth bricks ranges from 1600 kg/m³ for hand-made earth bricks to 2100 kg/m³ for extruded bricks. The strength can be estimated on site with the help of a rebound hammer but as it is usually straightforward to remove one or two bricks, it is possible to test the compressive strength of the bricks more accurately in the materials laboratory. The strength of historical hand-made earth bricks typically lies between 1 and 2 N/mm², for extruded bricks up to 3 N/mm².

In contrast to the raw mass of earth bricks, the historical earth masonry mortar was often leaned down. Sand was used most commonly, but short pieces of chopped straw or flax fibres were also used (figure 10.29). The density of the mortar typically lies between 1500 and 1800 kg/m². The compressive strength, and where desired the tensile bending strength, can be measured by replasticising the mass and testing in the same way as modern earth mortars. The strength usually lies in the region of 1 to 2 N/mm². If earth bricks were laid in lime mortar, a strength of 1 to 2 N/mm² can also be used.

### 10.2.3.2 The construction of earth brick buildings

Numerous characteristics of earth brick construction echo those described earlier for weller earth construction and are not detailed here anew. The following description instead details the areas that differ from weller construction.

Earth brick construction was used for single and two-storey residential buildings as well as for smaller barns. As with weller earth construction, it is not uncommon to find

Fig. 10.30 : Mixed brick and earth brick masonry on a gable wall (alternate courses of earth brick visible as lighter stripes)

the combination of a solid ground floor and lighter timber-frame upper storey. In addition, earth bricks were often used in villages and small towns to build firewalls between rows of timber buildings. Many are still evident today. Due to the low thickness of earth brick walls, bricks were also used for the internal walls of weller and rammed earth constructions. Earth bricks were likewise used for filling the triangular gable of other buildings as weller or rammed earth construction techniques were not practical or rational for constructing such shapes.

In earth brick buildings one can observe a gradual thinning of the wall thickness with each storey, more so than in weller and rammed earth construction. The wall thickness of the ground floor of a two-storey building, as well as for barns, was about 50 cm, on the upper storey, just under 40 cm. Loadbearing internal walls are seldom thicker than 25 cm.

Older earth brick buildings were given a two-coat external render with an earth undercoat plaster and a thin lime plaster facing render. After about 1870 buildings were increasingly rendered with one or two coats of lime render only. It was not typical to scratch out the mortar joints prior to rendering.

In the period as brick started to be come an affordable but still comparatively expensive building materials (approx. 1870 - 1900), many buildings were made of a combination of earth and ceramic brick masonry. The outer face of the external walls was made of brick, while the inner face of the external walls, as well as all the internal walls were made with earth bricks. In a two brick thick wall (in today's brick format: 49 cm), the proportion of fired brick in the external wall lay by 38 %, for a one-and-a-half brick thick wall 50 % and for a single brick thick wall 75 % (figure 10.30). In many cases the

Fig. 10.31 a and b : Mixed masonry and stone facing of an earth brick wall

earth brick content of such walls is only discovered during conversion works. If the loads applied to such walls change as a result of conversion works or a change of use, the relevant building elements may be subject to slight deformation, and therefore crack formation, as a result of the different elasticity of the materials.

Examples of mixed masonry walls made of stone and earth in Germany are most commonly found in the northern states of Brandenburg and Mecklenburg-Vorpommern. Natural stone was typically used on the representative façade or as façade cladding (figures 10.31 a and b). Most of these buildings date from between 1830 and 1860.

### 10.2.4 Damages and repair of massive earth constructions

"The buildings in the hamlets round and about Altenburg that are built of earth with a two foot high stone base have stood some 200 years and will, judging by how they look, stand at least as long again." [Geinitz, 1822]. The majority of the existing stock of monolithic earth constructions is more than 150 years old and the oldest known surviving example was built some 350 years ago. Monolithic earth construction can therefore be seen as a durable way of building in the Central European climate. Nevertheless the majority of these buildings also exhibit significant building damages. In almost all cases these can be attributed to constructions that are insufficiently adapted to new uses and environmental conditions as well as poor maintenance. The main problem – which the author of the above quotation already foresaw – is rising damp. Alongside the negative effects on indoor room climate and aesthetic appearance, this has led to many buildings suffering structural problems. Very few buildings are entirely free of

damages. The following patterns of damage can often be observed, listed here according to the frequency of incidence:
- Material damage and reduced cross-section as a result of rising damp
- Loosening of surface layers as a result of weathering
- Render and plaster damage due to age and insufficient plaster adhesion
- Cracks of different origins
- Pest infestation
- Washed out sections as a result of defective roof coverings
- Moisture damage as a result of vapour barriers

Almost all of the above are not specific to earth building and apply equally to structures made of other materials such as brick or stone (figures 10.32 a and b). Due to the different material properties of earth, however, damages in earth buildings are often more pronounced than in buildings made of other materials. The primary reasons for this are the water-solubility of the binding characteristic of earth and its low strength. All damages – with the exception of specific kinds of cracks – can be directly or indirectly attributed to the effects of moisture. The different patterns of damage are often present in different combinations.

10.2.4.1 Material damage and reduced cross-section as a result of rising damp

The main problem in the conservation of historical monolithic earth constructions is rising damp and the damaging salts transported by the moisture.

Earth walls affected by rising damp exhibit increased moisture content and a deconsolidation of the material of the wall from the wall surface inwards. Salt efflorescence cannot be seen on the surface but if one scratches away loose material near the sur-

Fig. 10.32 a and b : Damage resulting from rising damp on a brick wall and a rammed earth wall

face one can often see the crystallised salts in the material with the naked eye. If the degree of deconsolidation is relatively advanced, the material breaks away and the cross-section of the wall is reduced (figure 10.33). A permanent crusty coat of crystallised salts does not form on the surface.

The cause of the damage lies in the use of capillary conductive mortars in the foundation and plinth area combined with the lack of a properly functional damp proof course and a corresponding amount of moisture in the ground.

The wall material is most susceptible to damage from rising damp and the soluble salts contained therein near the wall's surface in the evaporation zone. The pressure exerted by crystallisation and hydration of the salts as well as of water freezing and thawing destroys the cohesive structure of the material. The resistance of a material to such processes is dependent largely on its pore volume and tensile strength. Compared with other materials, earth as a building material has a low tensile strength and is therefore particularly susceptible. In addition, the high capillary conductivity of earth means that large quantities of moisture can evaporate via the surface of the material. This results in a large moisture concentration gradient, which accelerates the capillary transport and with it the progression of damage. In earth building materials that contain straw, as used for example in weller construction, an increased moisture content in the wall can lead to humification of the straw. Where this occurs, the reinforcing effect of the straw is lost and the strength of the material sinks. As straw decomposes it produces humic acids which bring about changes in the clay minerals. In solid earth constructions, increased acidity, depending on the kind of acid, leads firstly to a transformation of the lime contained in the earth into expansive minerals and secondly to cation exchange in the intermediate layers of the sheet silicate in turn reducing the cohesive strength while increasing the swelling capacity of the minerals.

Fig. 10.33: Even very slight ground moisture is sufficient to create structural problems. Desert fort in the United Arab Emirates, 2007

The leads to further deconsolidation of the grain skeleton, not only on the surface but also within the cross-section of the wall. The combination of this material disintegration process and the fact that the compressive strength of earth is also a factor of its moisture content, the structural stability of the remaining cross-section becomes increasingly compromised. Measurements taken on numerous buildings showed a moisture level of 8% by mass in the remaining cross-section. The practical moisture content is then somewhere between 0.8 and 2%. In a series of tests, a mois-

Fig. 10.34: Introduction of a horizontal damp proof course including replacement of wall material with brick and earth brick masonry. Weller earth building in Steinstücken/Berlin

ture level of 8 % was shown to cause a 70 % reduction in compressive strength compared with the compressive strength under normal climatic conditions.

The elimination of rising damp and the repair of damages resulting from it is one of the most difficult building conservation tasks, not just for massive earth buildings. A prerequisite for undertaking successful repairs is a well-grounded concept based on a survey of the existing building and its pattern of damages as well as an analysis of the damages based on the existing material, the construction and not least the future use of the building. Many historical earth constructions exhibit evidence of previous improper repairs. In most cases the reduced cross-section is augmented with patches of brickwork and covered with diffusion-resistant insulation and plaster. Only rarely is the cause of the damage eliminated. Such measures lead to increased levels of moisture in the covered section of the wall and cause the moisture horizon to rise higher in the wall. The extent of damage is therefore increased still further.

A sustainable solution can often only be achieved through the partial replacement of material combined with the introduction of damp proofing layers. The precise approach depends on the specific conditions of each individual case. Mechanical approaches are preferred. The replacement of earth material is often necessary anyway as earth building materials containing soluble salts cannot be desalinated. The amount of material moisture required for such procedures exceeds what earth building materials are able to contain without destroying them.

If the decision is made to replace part of the wall material, the removed material should be replaced and augmented by earth brick masonry regardless of the construction method originally used. The mechanical properties of earth brick masonry

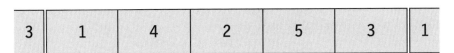

Fig. 10.35: Diagram showing the working process when replacing sections of walling

are compatible with those of other solid earth construction and the degree of shrinkage of earth brick masonry is lower than that of weller or rammed earth. The replacement procedure involves the metre for metre removal and replacement of material following a staggered pattern: after replacing a section of wall, the next section to be replaced may not be adjacent but should be offset to ensure sufficiently stability (see figure 10.35). The height of the damp proof course should be chosen depending on the height of the floor inside, and of the ground outside. Where necessary two damp proof courses may be required. Earth building materials may not be used beneath the damp proof course as the moisture level beneath the damp proofing will be much greater. Above the upper damp proof course (on the outside at least), a layer of water-impervious bricks should be laid. Special care should be taken that the upper junction to the existing structure bonds structurally with the wall above. This junction should be completed only once the process of shrinkage of the replacement masonry, which is comparatively slight, is largely complete. The junction should be closed with cement-based swelling mortar to avoid the risk of settlement.

### 10.2.4.2 Damages to plaster and render, weathering and washing out

The cause of damages to plaster and render lies predominantly in the use of mortars that are too hard. This causes large-scale separation of the render and raises the moisture horizon in the case of rising damp by reducing the ability of the wall to dissipate moisture by evaporation. The weathering resistance of unrendered massive earth surfaces is generally good due to the straw present in almost all historical earth mixtures. As such, a lack of render was better than an improper render as the degree of surface erosion, especially of weller earth constructions, is surprisingly small. On several weller earth barns dating back to the mid-19th century, weathering of no more than 50 mm was discovered on the façade exposed to the weather, although the next 20 to 50 mm did exhibit a loss of strength (figures 10.36 a and b). The processes that cause damage to the material structure are comparable to those of rising damp, with frost-thaw cycles playing a larger role than salts.

In contrast to the relatively good resistance to exposure to rainwater, running water as a result of defective roof coverings can lead to considerable erosion of the surface material (figure 10.37). The edges of eroded channels typically contain salts and should be removed. When replacing these sections with earth brick masonry, more material must be cut away to create rectangular bearing surfaces and a lateral bond for the brickwork. An additional bond to the existing wall can be established through the inclusion of metal wall anchor strips.

The most important aspects to consider when applying a new render is to sufficiently prepare the weathered substrate and to choose a rendering system that is compatible with the substrate.

The following steps should be taken when preparing a weathered surface for receiving render:

- Removal of any disintegrated earth material followed by at least a further 3 cm of undamaged materials using a masonry hammer
- Roughening of the surface with a masonry hammer
- Brush down all loose material with a medium-stiff wire brush
- Wet the surface sufficiently
- Work in a slurry of sufficiently cohesive earth and angular sand with a hard brush

All material removed from the weathered surface should not be reused as it is generally contaminated with critical concentrations of salts (section 2.4.2.5). In addition, complex mineral conversions have taken place while exposed to the weather which usually lead to much greater shrinkage behaviour. The latter can be compensated for through the addition of additives but both phenomena typically occur together. Undamaged earth material can, however, be replasticised and reused. If in doubt, a salt analysis should be undertaken.

The pre-prepared surface is then coated with one or more equalising layers of earth mortar containing angular sand and chopped straw. A single layer should not be thicker than 2 - 3 cm. Where there are deeper holes, wooden nails should be hammered in to provide an additional anchor for the plaster-like application. Holes of up to about 7 cm

Fig. 10.36 a and b: Weathering crust on a weller earth surface exposed to the weather (a) and the undamaged material after removal of loose material (b)

Fig. 10.37:
Erosion channel resulting from a defective eaves covering

Fig. 10.38:
Replacement of a heavily weathered earth brick wall with a render-like application and earth brick masonry (Muwaiji Fort, Al-Ain, UAE)

should be filled with layers of the earth material. Before a new layer is applied, the first layer should have dried to the point where no further cracks form. Holes deeper than about 7 cm can be filled with earth bricks cut to size. The kind and density of brick should be chosen to correspond as best possible to the composition and density of the existing wall, and should be declared as Usage Class I by the manufacturer (figure 10.38).

If the substrate has been prepared as described above, the topcoat render can be applied. According to the rules of the trade, external renders should be more vapour permeable and elastic than the substrate. This also applies to the underlying layers. Monolithic earth substrates have a very low vapour diffusion resistance and high elasticity. To adhere to the above rule of thumb it is necessary to use a straw-fibre reinforced earth plaster. This presents the least risk of flaky separation from the substrate as a result of differential thermal expansion and contraction or vibrations. The *Lehmbau Regeln* restrict the use of earth plasters to surfaces not subject to driving rain but note that when skilfully undertaken, external earth renders can also demonstrate good weathering resistance. In recent years, the addition of cow dung to the earth render has successfully been used in the renovation of weller earth buildings. The surfaces were only partially coated, typically using a lime-casein paint.

For rendering historical earth buildings, lime plasters with a strength $\leq 1$ N/mm² are particularly suitable. Without hydraulic additives, they must be painted with a suitable paint to achieve an adequately frost-resistant surface. Coatings with an $S_d$-value < 0.1 m and a w-value of 0.3 to 1.0 kg/m²h$^{0.5}$ are recommended.

Since around 1990 good results have been obtained using air-hardening lime renders with hydraulic admixtures. Pozzolanic admixtures (for example in the form of powdered brick) are the most common hydraulic factors used in such cases. Selected commercially-available, factory-produced ready-mix mortars from this plaster group with a strength of 2 - 3.5 N/mm² and a water vapour diffusion resistance coefficient of 10/15 are compatible with the plaster substrate.

Lime-cement mortars and mortars made of very hydraulic limes are not suitable due to their high strength and low elasticity. Synthetic-resin-bound renders should not be used at all. While they are sufficiently elastic their vapour diffusion resistance is too high.

Plaster laths are generally not necessary as the substrate is typically sufficiently roughened during preparation. Reinforcement mesh can, however, be used to bridge changes of material and crack repairs.

### 10.2.4.3 Cracks

Settlement cracks in massive earth constructions are mostly found in barns and only rarely in residential buildings. Due to the comparatively long length of walls in barns, the probability of differential settlement as a result of too shallow foundations is more likely.

In addition, the lack of a ring beam can lead to gaping cracks arising. These can occur in the middle of walls (figure 10.39) as well as at the corners and junction areas of walls (figures 10.40 a and b).

A further typical pattern of cracking results from the use of materials of different elasticity in the direction of load dissipation. For example, the jambs of walls and entrance doorways are often lined with brickwork. The stronger degree of shrinkage and creep of the earth leads to settlement cracks forming (figure 10.14).

When planning renovation measures, it is important to first ascertain whether the cracks and the building are no longer moving and that the spatial stability is guaranteed. If this is not the case, a concept must be elaborated for reestablishing the spatial stability for each affected building, including a precise analysis of all the influencing factors. In the simplest case, the closing of a crack may be sufficient. Similarly the insertion of ties into the ceiling level or ceiling slab is an often simple but effective measure. More complex measures including needling or the later insertion of ring anchors or ring beams. Such elements are typically made of reinforced concrete cast in-situ. Ring beams made of wood are more difficult to realise as the connection between the ring beam and the fabric of the existing building usually requires the additional insertion of multiple complex anchorages.

Needling is used to bridge cracks. To insert the needles, approximately 3 cm wide and 20 cm deep channels are cut into the wall at right angles to the crack direction. After blowing out and wetting the channel, a reinforcement needle is inserted and bedded in enveloping mortar.

Fig. 10.39 : Settlement crack as a result of differential settlement, Bobbau/ Saxony-Anhalt

Unlike anchors where the tensile force pulling apart the crack is resisted by anchor plates, needles resist movement along their entire length through friction. Because the friction between the needles (typically ribbed glass fibre or stainless steel reinforcement rods) and earth is very low, the channels are filled with a firm mortar (trass-lime or cement mortar) that envelopes the needles. The stresses resisted by the reinforcement rod are transferred via the comparatively large surface area of the mortar into the earth. At present, relatively little is known about the necessary diameter and length of needles or the stresses they need to sustain.

The repair of cracks at corners requires a combination of the needling and anchorage principles. Because at the corners of walls there is not enough wall depth in one direction for the needle, one end of the needle must be equipped with an anchor plate (figure 10.41).

Fig. 10.40 a and b : Settlement and shrinkage craces at the corner of a chalk-pisé house (rammed earth with additional chalk content) in Weferlingen an der Aller / Saxony-Anhalt, 2008

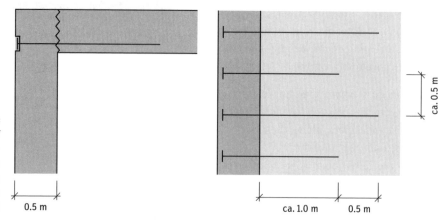

Fig. 10.41:
Securing the corner of a building with a combination of needles and anchors. Diagram showing plan (left) and cross-section (right)

### 10.2.4.4 Pest infestation

Where monolithic earth constructions suffer from rising damp, the wall structure can be penetrated by mice, rats and ants. This can weaken the loadbearing cross-section considerably. The walls should be protected against rising damp by the introduction of a damp proof course and the affected areas replaced with new earth material.

The south and west façades of monolithic earth walls that have not been plastered may be colonised by solitary bees, even when not affect by rising damp (figures 10.42 a and b). Solitary bees live alone in ca. 10 cm long vein-like tunnels with a round bulging end and are a protected species in Germany. The repair of affected parts of a building can lead to a collision of interests between structural concerns and the policies of building conservation and nature preservation. The tunnels themselves are not a problem for the structural stability of the wall. More problematic, however, is the loss of building fabric caused by birds picking away at the earth in an attempt to get at the bee's brood (figure 10.42b).

### 10.2.5 Thermal insulation

Due to the different bulk densities and wall thicknesses of the external walls of monolithic earth constructions, they should be considered individually with regard to thermal insulation. A positive aspect of all construction methods is the high thermal retention capacity of such walls due to their thickness and solidity, a product of their material composition and construction method. Weller earth walls have a much lower density than earth brick and rammed earth walls. The thermal conductivity coefficient of historical weller earth buildings is calculated as being U ≈ 0.8 W/m²K, for earth brick constructions ≈ 1.2 W/m²K, and for rammed earth constructions ≈ 1.3 W/m²K. Weller construction was therefore always regarded as being better in terms of thermal per-

Fig. 10.42 a and b : Solitary bee entrance points in a weller earth wall

formance: "Earth weller walls protect against frost better than stone walls" [Geinitz, 1822].

The German Energy Conservation Regulations, the EnEV, requires that heated buildings must be insulated when a façade is renovated. Exceptions and relaxations are possible where this requirement conflicts with conservation aspects, building material properties or construction, or is not economically feasible. Building material and construction concerns are not a problem as compatible vapour-permeable insulation materials are available. The necessary insulation thickness must be chosen with respect to the existing construction. Typically 8 - 12 cm of insulation is required to fulfil the minimum requirement. Before insulation measures are undertaken, the walls must be repaired and measures taken to ensure the wall remains dry.

A moderate level of internal insulation is possible (see chapter 7).

## 10.2.6 Building material and building element properties

### 10.2.6.1 Mechanical properties

Table 10.1 : The spectrum of mechanical properties of historical monolithic earth building materials

| Building material | | | Weller earth | Rammed earth | Earth bricks |
|---|---|---|---|---|---|
| Bulk density | r | kg/m³ | 1400 – 1700 | 1600 – 2200 | 1600 – 2100 |
| Compressive strength | b | N/mm² | 0.6 – 1.3 | 1.5 – 3.0 | 1.0 – 3.0 |
| Modulus of elasticity | E | N/mm² | 250 – 400 | 500 – 800 | 400 – 700 |

## 10.2.6.2 Selected physical properties

Table 10.2 : The spectrum of physical properties of historical monolithic earth building materials

| Building material | | | Weller earth | Rammed earth | Earth bricks |
|---|---|---|---|---|---|
| Bulk density | r | kg/m³ | 1400–1700 | 1600–2200 | 1600–2100 |
| Thermal conductivity | l | W/mK | 0.6–0.8 | 0.7–1.4 | 0.7–1.2 |
| Water-vapour diffusion resistance coefficient | m | N/mm² | 5/10 | 5/10 | 5/10 |

The fire performance of all building materials conforms to building material class A.

Table 10.3 : U-values of typical historical monolithic earth wall cross-sections

| Building element | | | Weller wall $\rho$ = 1500 kg/m³, d = 60 cm, plastered on both sides | Rammed earth w. $\rho$ = 1900 kg/m³, d = 50 cm, plastered on both sides | Earth brick wall $\rho$ = 1700 kg/m³, d = 50 cm, plastered on both sides |
|---|---|---|---|---|---|
| Thermal conductivity coefficient | U | W/m²K | 0.85 | 1.35 | 1.2 |

Table 10.4 : Necessary insulation thicknesses for the external insulation of typical historical monolithic earth wall cross-sections with softwood fibreboard insulation ($\lambda$ = 0.045) according to EnEV 2009

| Building element | | | Weller wall $\rho$ = 1500 kg/m³, d = 60 cm, plastered on both sides | Rammed earth w. $\rho$ = 1900 kg/m³, d = 50 cm, plastered on both sides | Earth brick wall $\rho$ = 1700 kg/m³, d = 50 cm, plastered on both sides |
|---|---|---|---|---|---|
| Thickness of insulation layer | d | cm | 14 | 16 | 15 |

Table 10.5 : Weighted sound reduction index $R'_w$ acc. to DIN 4109, supplement 1, for single-leaf rigid walls

| Building element | | | Weller wall $\rho$ = 1500 kg/m³, d = 60 cm, plastered on both sides | Rammed earth w. $\rho$ = 1900 kg/m³, d = 50 cm, plastered on both sides | Earth brick wall $\rho$ = 1700 kg/m³, d = 50 cm, plastered on both sides |
|---|---|---|---|---|---|
| Mass per unit area | | Kg/m² | 950 | 1000 | 900 |
| Weighted sound reduction index | $R'_w$ | dB (A) | 57* | 57* | 57* |

*Applies for single-leaf rigid walls with a mass per unit area of more than 580 kg/m².

Fig. 10.43 : Wielandhaus in Biberach/Riss

## 10.3 Timber-frame panel infill

In Central Europe timber-frame construction, in which a structural timber frame encloses panels filled with a walling material, developed alongside masonry and other means of solid wall construction. Among the oldest surviving timber-frame constructions in Germany are the Gotische Haus on the Römer 2-6 in Limburg (1289) and the post-and-beam building in the Wordgasse 3 in Quedlinburg (early 14th century). The very wide framework of the Wielandhaus in Biberach/Riss (1318/19) is an indication of its origins from the early period of timber construction (figure 10.43). It resembles archaic constructions where the structure was formed by posts placed far part.

In many regions stone construction gradually became the predominant building method but in other regions timber-frame construction continued to develop and in still other regions both methods developed in parallel alongside one another.

The availability of suitable wood for construction was an essential prerequisite for the development of a culture of timber construction. Many of the regions in Germany that are characterised by timber-frame construction are located in the densely wooded

Fig. 10.44:
Despite settlement of the timber-frame, the panel infill is still fundamentally intact

highlands of central and southern Germany. In regions where wood was not as plentiful, monolithic earth construction was more widespread.

While the timber framework of timber-frame construction has been the subject of extensive research for many years, historians have not until recently devoted much attention to the architecturally less remarkable aspect of panel infill.

Earth building materials were used in timber-frame construction as panel infill to close off the surfaces of rooms. A primary reason for the use of earth in timber-frame construction was its availability. Earth and other materials for the panel infill needed to be easy to procure without undue cost or effort. A further key reason for the use of earth was the ease with which it can be formed and worked.

In past centuries the work of creating the panels of timber-frame constructions was not necessarily undertaken by specially trained craftsmen. It was probably undertaken by day labourers and travelling workmen, perhaps also by families or local communities. As such, panel construction underwent little in the way of ongoing technical development [Figgemeier, 1994]. The continuity of the materials and techniques used over many hundreds of years nevertheless indicates that their use was both conscious and deliberate and based on many years of experience.

To our eyes, the old methods of creating panel infill do not seem very optimised but their simplicity meant that they were particularly adaptable to the specific conditions of historic timber-frame constructions. In many regions, irregularly formed beam cross-sections and differently shaped panel geometries are characteristic for timber-frame constructions.

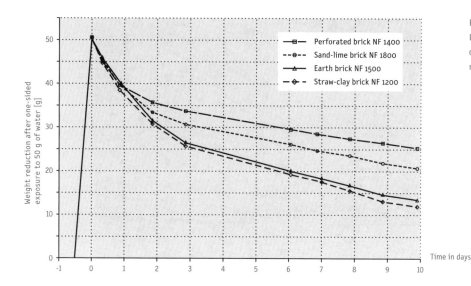

Fig. 10.45:
Drying behaviour of different panel infill materials

An important technical property of old earth panel infill material is its resilience. Compared with other building materials, earth has a comparatively low strength. The building methods used for the panel infill also permit the panel to respond to movement within the timber frame, a typical and often largely mechanical phenomenon. This adaptability of the infill core can be seen quite clearly in figure 10.44. Here the flexibility of the panel allows it to give, whereas a rigid infill of hard bricks could instead damage the structural framework.

The relationship between the timber-frame construction, the geometry of the panel infill and the panel infill technique employed is central to understanding the structure as a whole. The fact that these conditioned each other and developed alongside one another shows that timber-frame walls can be regarded in technical as well as cultural terms as a whole. Conservation practice likewise considers the timber framework and the panel infill as integral components of the same traditional structural system. The obvious congruence between the historical natural building materials earth and wood is a further reason why owners of historical timber-frame constructions often express the wish to maintain the substance of their building and to augment it with like materials. It is likewise impressive to see just how long the old building materials and techniques have survived.

The primary requirement for panel infill structures that are exposed to the weather is the ability to dry out quickly after unavoidable moisture ingress. For this reason, it is important to ensure that the materials used exhibit good capillary conduction [WTA, 1999]. The earth building materials used for historical panel infill have good

Fig. 10.46 : Panel infill of gathered stones bedded in earth mortar in Weimar (Thuringia)

capillary conductivity and, compared with other materials, dry out very quickly (figure 10.45).

### 10.3.1 Panel infill techniques

The proper repair of earth infill panels requires an understanding of old construction methods, most of which are simple but no longer common knowledge. The following section describes the three most important methods of creating panel infill with earth building materials along with the materials required. Two of these use either a grate or latticework made of wood as the supporting construction of the panel infill: *wattle with straw-clay daub* and *staves with straw-clay*. The third technique uses *earth block masonry* for the infill core without a wooden construction.

A further technique is to fill the panel with gathered stones (typically stones removed from open fields so that they do not obstruct agricultural machinery) embedded in earth mortar, with or without the help of shuttering. The technique is similar to that of stone walling and is therefore not discussed in further detail here (figure 10.46).

#### 10.3.1.1 Wattle with straw-clay daub

The construction of wattles with straw-clay daub begins with the insertion of *staves* wedged between the beams. The ends of the staves are set in *grooves* cut into the side flanks of the timber structural members with a hammer, axe or similar. In many cases only one side exhibits a continuous groove while the opposite side has only short slots or bored holes (figure 10.47). The depth of the grooves or notches is usually 1 - 2 cm, the width 2 - 3 cm. The distance between the groove and external face of the timber-frame structure was often chosen so that there was space enough for the wattle, and a 1 cm thick daub covering of straw-clay. The final facing coat of finishing render was typically applied in a very thin layer, if at all.

Staves were long pieces of wood such as a batten or stick wedged between the beams of the timber framework. The staves are usually made of the same wood as the primary construction, but other kinds of woods could also be used. In most cases the staves were produced by splitting a section of wood and then rounding off its edges to form a round or elliptical cross-section. Pieces of roundwood were also used, then with their full cross-section. The sapwood was not always removed but each stave had a sufficiently stable core. Before insertion the ends of the staves were pointed on one or both sides, occasionally all round.

Staves were usually inserted vertically, but the horizontal arrangement of staves was also known. In triangular or trapezoidal panels, the staves could fan out to fill the panel. Another alternative was to space out the wattle to a fan shape (figure 10.48). In general the staves usually bridged the shorter distance of the panel dimensions.

The staves were sprung between the beams so that they were held firmly in place. The spacing between the staves was chosen to be far enough apart to comfortably insert the wattle rods but not so far apart that the wattle could wobble. Very often larger panels were divided into two sections for easier construction of the wattle. Due to the rounded off profile and polygonal cross-section of the staves, the outermost stave only touched the beam at certain points. In many cases a space was left between the timber framework and the outermost stave.

For the *wattle*, flexible branches or *withies* were chosen that were easy to weave between the grate of staves. At the same time, they had to be sufficiently stable and as resistant as possible to wetting. In many regions straight sections of fast-growing willow rods were used, but hazel or other branches were also commonly used. To make them easier to work, the rods or branches were split twice or three times down the middle. The protruding ends of the wattle branches were usually sliced off diagonally with an axe (figure 10.49).

The branches are woven alternately in front and behind the staves to form a more or less rectangular latticework. A criss-cross weave was also widely used (figure 10.50). The rounded edges of the staves made it easier to weave the branches around them and avoided breaking them. The distance between the wattle rods was only as large as it needed to be to press the malleable earth daub between them. Depending on the constitution of the daub, the distance ranged between the thickness of a finger and that of the middle of the hand. For very fine straw-clay mixtures, the distance between the rods could be made smaller and sometimes the branches even touched.

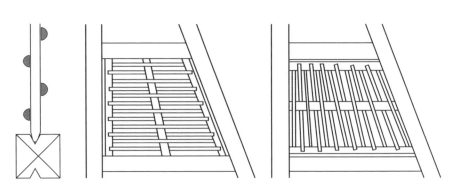

Fig. 10.47 :
Cross-section showing staves and wattle

Fig. 10.48 :
Either the staves or wattle were fanned out to fit irregularly shaped panels

The *soil for the daub* was often sourced directly from the site or a clay or earth pit in the vicinity. These were often located at particular locations and made available for use by the whole village. Many old street and land-parcel names testify to their former location. The earth mixtures used for daub typically had a more lean consistency.

*Straw* was added to the earth mixture to improve the stability and weathering resistance of the panel as well as to lean down the mixture. Like earth, straw was readily available. Barley straw and rye straw were most commonly used, sometimes straw from older kinds of cereals, usually of cut length as harvested or as long chopped straw. The straw content was often quite high with mixtures attaining a bulk densities of around 1400 kg/m³. Even lower bulk densities of ≤ 1200 kg/m³ (light straw-clay) were sometimes also used [Volhard, 1992]. Mixtures with shorter and finer plant fibres were often used for the finer topcoat layers on the inside and outside.

Fig. 10.49 :
The panel of a timber-frame house in Saxony

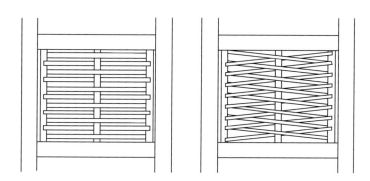

Fig. 10.50 : Perpendicular and diagonal weave patterns for the wattle

When making the *straw-clay daub*, water was probably first added to the soil or earth mixture before mixing in straw of different lengths [Figgemeier, 1994]. Animals were often used to tread the mixture.

The addition of cow dung helped improve the workability of the straw-clay. It also improved the strength and weathering resistance of the panel.

*Tempering* was a further means of improving the workability and properties of the mixture. By allowing the mixture to stand in a malleable consistency, microorganisms within the mixture formed lubricating films that help the clay minerals slide over one another, increasing the plasticity and final strength of the mixture. The earth mixture was broken down more and the straw softer, suppler and easier to mould as a result of water uptake.

The application of the first and thickest layer of straw-clay was usually undertaken on the inner face of external walls. Tests undertaken on the *Gotische Haus* in Limburg showed that large lumps of daub of 25 × 25 cm made of earth and long-stem straw had been applied. The thickness of the lumps usually corresponded approximately to the distance between the branches, although some lumps were up to 9 cm thick. One end was hung between the branches and then pressed into the web of branches below. The pattern of application indicates that the material was applied with bare hands [Volhard, 1992].

The second application was undertaken from the outside face of the external wall once the first layer had dried. The thickness of the layer was thinner than the first application, the material typically somewhat finer and more like a *straw-clay plaster* mixture. Both applications were undertaken by hand but may have been beaten or smeared into place with a simple tool such as a flat piece of wood.

Fig. 10.52 :
Staves wrapped with 'ropes' of straw-clay in parallel formation and fanned arrangement

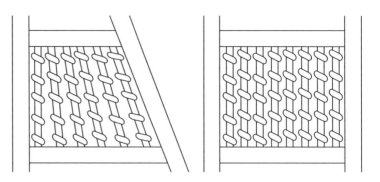

On the inner face, the first application of daub was generally simply smoothed over. In older buildings, and later too for simple building constructions, this sufficed as the final surface finish.

10.3.1.2 Staves with straw-clay

With this technique the staves are arranged closer together to form a lath-like grate. The grooves in the beams were continuous and their dimensions and location in the cross-section of the timber beam followed the same principle as described previously for *wattle with straw-clay daub*. The staves were mostly arranged vertically. The distance between the staves was chosen to be large enough to facilitate the subsequent work stages. Typically the space was about as wide as a hand thickness.

Fig. 10.51 :
The wall of a timber-frame building in the state of Brandenburg

In Germany, this panel infill method was more common in East Germany where softwood timber-frame construction predominated. The staves were usually made of the same wood as the primary structure (see above). In regions where softwood timber-frame construction prevailed, the good splitting characteristic of long-fibre woods may be a reason for their use, due to the easier fabrication of large numbers of staves. This could be a possible explanation for the increased predominance of this method in East Germany.

In East Germany, one can also observe that the panel infill of older timber-frame buildings were often constructed as wattle and daub while those of buildings from the 19th century onwards used staves and straw-clay. The later availability of better splitting tools may explain this shift in the construction method.

Three different but related means were used to cover the staves: *staves with straw-clay daub*, *staves wound with 'ropes' of straw-clay* and *wound earth reels*. The techniques are all similar in principle and the dividing line between them blurred. Walls were not always consistently handled in one or other technique: sometimes the space between the staves were smeared with straw-clay daub, sometimes parts of, or more commonly whole panels, were wound with 'ropes' of straw-clay (figure 10.51).

*Staves with straw-clay daub* simply received a coat of malleable straw-clay daub that was pressed into and through the spaces between the staves in a process almost identical to that of wattle and daub.

To improve the adhesion of the straw-clay mixture, the staves were sometimes wound with *straw-clay 'ropes'* (figure 10.52). For this the long-stem straw as harvested from the field was mixed with a malleable mass of earth or alternatively softened directly in a malleable earth mixture or earth slurry. After a while left to soften, the earth-soaked straw became more supple and could be formed into either simple bundles of strands or entwined strands to simple ropes and then wound around the staves. The thickness of the straw-clay 'ropes' could be a few centimetres or as thick as an arm. In many cases there are only three to five windings around a stave. A variant of this was *staves with interwoven straw-clay 'ropes'*. The ropes were made as described above but instead of winding them around the staves they were interwoven between the staves.

Where the staves were fully enclosed by the windings of straw-clay 'rope', the resulting element was known as a *Wickelstake* or *wound earth reel*. These were most commonly used as a ceiling infill material arranged between timber beams (section 10.4). The straw-clay strands were either wound around the stave in a helical screw pattern or wound as separate rings around the staves. A further method involved winding a mat of straw-clay almost as wide as the stave was long around the stave.

The final covering of the internal and external face of the panel infill core followed the same principle as described earlier for panel infill with *wattle and straw-clay daub*.

Fig. 10.53: Historical panel infill with earth brick masonry (Taunus)

### 10.3.1.3 Panel infill with earth brick masonry

Panel infill was also made of earth brick masonry (figure 10.53). The straw content suggests that special mixtures were created for this purpose. The format of the bricks typically lay somewhere between today's DF (thin format) and NF (normal format), sometimes slightly larger. Bulk densities of between 1400 and 1800 kg/m$^3$ were typical.

To secure the often half-brick thick masonry panels against falling out, thin split branches or battens were nailed to the inward-facing flanks of the timber members of the timber-frame construction. In later examples long nails were also used hammered into the sides of the timber posts so that they projected into the mortar joints of the courses of brickwork.

The brickwork itself was laid using the normal masonry techniques and bedded in sufficiently leaned-down earth mortar, sometimes in lime mortar.

The panels were rendered with earth or lime mortar. On walls protected from driving rain, examples of fair-faced earth brickwork arranged flush with the face of the timber frame have also survived.

### 10.3.1.4 Internal and external facing coats

The application of layers of *internal plaster* over the entire wall surface, covering both the timber frame as well as the panel was a significant step towards improving wind-proofing and thermal insulation. The photo shows a coat of plaster containing straw with stem lengths of up to 25 cm, coated with a fine-grain earth plaster reinforced with flax fibres and painted with a coat of limewash (figure 10.54).

Fig. 10.54 :
Internal facing plaster applied over the entire surface of the wall covering both panels and timber frame

*On the external face* a layer of straw-clay applied to the outside face of the panel infill or facing brickwork may have served as the facing layer, particularly in older timber-frame buildings or simple buildings. In most cases, however, the surface of the panel core was recessed behind the face of the timber frame by about ½ cm to leave enough space for a final coat of fine straw-clay render. The external render was also necessary to cover the shrinkage cracks in the surface of the infill core.

The straw-clay surface was often roughened to provide a mechanical key for subsequent layers of render or plaster. In some regions, one finds examples of wavy or diamond patterns scored in the surface, which were sometimes undertaken with remarkable regularity. Some of these patterns were analysed in buildings in Heuersdorf south of Leipzig in East Germany. Typically they were scored with a comb-like scratcher into the still-wet earth surface. The process was so simple to undertake before the earth dried that one sometimes also find such marks on surfaces that were never later plastered.

The scoring of the surface can cover parts of the panel or also the entire surface. Where such marks were made, they always ran around the edges of the panel parallel to the timber members of the framework, where the subsequent plaster or render would need to adhere particularly well. The design and regularity of the scoring patterns may also have served a certain decorative touch, but their relevance as religious or mythical symbols should not be overestimated (figures 10.55 and 10.56) [Scholz, 1998]. An alternative approach to scoring the surface was to roughen the surface by imprinting the surface with a pricked pattern of holes. The pricking of patterns of holes with simple fork-like tools was also widespread (figure 10.57).

Fig. 10.55 (left): Diamond pattern scored into a straw-clay infill panel (Saxony)

Fig. 10.56 (right): Wavy scratch pattern scored into a straw-clay infill panel (Saxony)

To protect earth renders against the weather, the top coat of render was often coated with thin layers of diluted limewash. In many regions, such coats were applied to both the panels as well as the timber beams. The coats were sometimes so numerous that they are easy to mistake for lime plaster.

The highlighting of the graphic effect of timber-frame constructions by painting the beams red or black and the plaster panels white became popular in many regions from the 19th century onwards. Before such time, the residents obviously saw no need to embellish the façade of timber-frame buildings.

The limewash could also be mixed with fine sand (figure 10.58). In the period after 1945, Niemeyer recommended using a 1:1 mixture of white hydrated lime and sand as a base for subsequent paintwork. The slurry was worked into the pre-wetted earth substrate with a hard sponged or felted float while allowing the earth to still show through [Niemeyer, 1946]. This produced a stabilised surface between the lime and the earth to which the following layers of lime paint or wash could adhere more properly.

Renders made of quicklime and coarser sand have also been commonly used for many years, becoming more and more prevalent in many regions from the mid-19th century onwards. Such layers of plaster were usually only a few millimetres thick and by today's standards only suitable for internal plasters. These thin applications were, however, lightweight and supple and have in many cases proven their resilience as external render over long periods.

Fig. 10.57 (left):
Holes made in a base coat with a four-pronged fork-like tool

Fig. 10.58 (right):
The remains of an old limewash with granular texture

## 10.3.2 The repair of external timber-frame panels

In general, all repairs should follow the principle of the existing panel as closely as possible. The following recommendations are based on many years of practical experience in the renovation of numerous timber-frame buildings [Breidenbach/Röhlen, 2008]. Possible methods for improving the thermal insulation are discussed in chapter 7 *Internal insulation with earth materials*. The necessary materials for repair works can be sourced from earth building material producers. The properties of the straw-clay mixture for the repair of panel infill should correspond as far as possible to the existing straw-clay mixture. One should not, for example, apply too rich mixtures to a lean, sandy substrate. Factory-mixed materials may therefore need adapting to the conditions on site. The manufacture of site-prepared mixtures of earth and additives is possible with the requisite experience. If material recovered from the old panel infill is to be reused, it first needs to be checked for contaminants such as salts, soot and wood-rotting fungi.

### 10.3.2.1 The repair of panels made of wattle & daub or staves & straw-clay

Panel infill made of earth materials can often be retained and repaired even when they are hundreds of years old. One should not be deterred by certain, by today's standards "unusual" building properties: loose staves can be fixed in place; insect holes may be old and unproblematic; straw-clay may be soft but it is firm enough for the purpose required.

Fig. 10.59:
Example of a badly damaged timber-frame panel

Diagram showing wattle and old straw-clay, 20 mm new straw-clay, 15 mm lime render (from right to left)

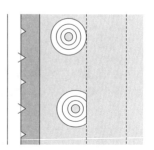

The replacement of defective staves or wattle usually entails a greater loss of existing substance. In such cases it is preferable to stabilise what still remains through the use of non-corrosive screws or other means of fixture.

Before beginning with repairs, any remaining plaster or previous repairs made of lime or cement mortar should be removed. Sections of earth that fall away on rubbing one's hand over them should be removed. If staves form a more or less closed surface, a plaster lath should be screwed to the staves to improve adhesion.

The choice of the desired plaster coating determines the kind of repair: on heavily weathered surfaces, which usually exhibit a fairly significant degree of damage to the earth material, a *two-layer application of lime render* of about 15 mm thick should be applied (figure 10.59). On surfaces not so exposed to the weather, where the facing surface is very often partially intact and the straw-clay only slightly damaged, a 3-5 mm thick *single coat of lime render*, as commonly used in historical buildings, is usually sufficient (figure 10.60).

For a two-coat application of lime render, a sufficiently rough plaster base is required. This can achieved by carefully roughening the freshly applied coat of straw-clay which requires a plaster layer of about 2 cm thick.

As described above, the forward face of the wattle or earth 'ropes' sometimes lies just 1 cm behind the front face of the timber beams. This necessarily means that the stra-clay and external render repair must bulge outwards beyond the surface of the timber beams. Towards the edge of the panel, the thickness of the straw-clay layer must be reduced so that the protruding surface can fall back to finish flush with the timber beam. This should only take place within the layer of straw-clay and ease off gently. Where

## 10.3 TIMBER-FRAME PANEL INFILL

Diagram showing wattle and old straw-clay, 3 mm lime render (from right to left)

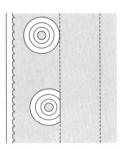

Fig. 10.60 : Example of an only slightly damaged timber-frame panel

space is limited, it may be necessary to remove a little of the old straw-clay material at the edge of the panel. The thickness of the final facing lime render must be uniform.

Before the application of the new straw-clay layer, the existing infill should be gently wetted. This binds any remaining dust and promotes a better bond between the old and new earth. The new earth is thrown on with a trowel or spread onto the panel. After it has begun to bond it should be roughened while still of a soft consistency with a render finishing scraper or similar (figure 10.61). The straw fibres that stick out of the surface help provide a mechanical key for the subsequent layer of render.

New layers of straw-clay must be allowed to dry free of frost. The surfaces should likewise be protected against heavy rain or splash water bouncing off scaffolding boards.

To prepare the surface to receive a thin single-coat of lime render, any holes in the layer beneath should be repaired with lean straw-clay or earth plaster mortar with straw fibres. The panel should then be gently wetted and thinned earth plaster mortar with straw fibres worked into the surface using a hard sponged or felted float.

### 10.3.2.2 The repair of panels made of earth brick masonry

The partial repair of earth brick masonry is undertaken analogue to the repair of masonry made of fired or synthetic bricks. The water solubility of the earth masonry mortar makes the job a little easier. Bricks and mortar should be used that correspond as far as possible to the material of the existing timber-frame panel. The stability of the panel can be weakened or even compromised entirely by the removal of individual bricks. For this reason one should carefully assess whether repairs are viable or whether a new panel infill would be better.

Fig. 10.61: Roughening of the repaired material with a render scraper (the edges of the panel will be roughened separately)

### 10.3.3 New panel infill

The renewal or augmentation of authentic historical constructions such as *wattle and daub* or *staves with straw-clay* may be necessary, for example in the conservation of historic buildings. In most other cases, new panel infill is typically undertaken using earth brick masonry for cost reasons. This has the added benefit of limiting the exposure of the timber frame to moisture and avoids the need to wait for long drying times.

#### 10.3.3.1 New panel infill with wattle and daub or staves and straw-clay

Before beginning with earth construction works, all carpentry work must be completed. Existing grooves may need to be deepened or widened. New grooves can be cut with a chisel or chainsaw at a sufficient distance from the front face of the timber framework, approx. 4-5 cm. The timber-frame members should not be given grooves along their entire length before insertion into the construction as this would result in channels running behind the junctions of timber members and problematic cavities in the construction.

Sawn battens can be used for new staves but the wood should correspond to the wood of the old staves and timber-frame construction. Staves from the existing structure that are in a good condition can be reused. The staves are pointed with an axe or saw and brought into place between the timber-frame members with a heavy hammer or the blunt end of an axe so that it sits tightly but does not force the timber members apart or split. A gap of approx. 2 cm should be left between the outermost stave and the side of the timber-frame panel.

For the wattle, seasoned and split lengths of willow branch are normally used. The sections are cut to length with a pair of shears. The branches should be chosen to be as thick as possible while still workable. Especially the uppermost couple of wattle rods must be sufficiently strong or else they risk sliding down when the fresh straw-clay mix is applied. The distance between the rods should be the thickness of a thumb or the middle of a hand. The application of the straw-clay daub should follow as soon as possible as the wattle starts to lose its clamping force as time passes.

If one wishes to use *staves* for the *straw-clay daub method* that are partially or completed wound with straw-clay rope, the method described in the section on wound or rolled earth reels for ceilings should be used (section 10.4).

For *straw-clay mixtures*, rye or ideally barley straw should be used. The pieces of straw should be left long at around 5 - 20 cm. The proportion of straw in the mixture should be high. The empirical value for the proportion of straw content given in the *Lehmbau Regeln* is 40 - 60 kg/m³. The richer the earth the larger the proportion of straw content can be. For better workability the material should be left to stand for several hours prior to use. The malleable mixture is usually slapped on vigorously from the outside with a plaster smoother or thrown and applied with a masonry trowel. To ensure that the entire wattle and staves are properly enveloped in the mixture, sufficient straw-clay must be pushed through the wattle. The distance between the straw-clay and the face of the timber frame should ideally be exactly equivalent to the thickness of the render to follow. The resulting surface is then roughened or indented with holes (see above) while still soft. After a while, the straw-clay material that has been pushed through to the inside is then pressed back onto the staves and the inner face completed with a thin more supple straw-clay mixture so that a closed surface results.

Good drying conditions must be ensured by allowing wind and draught to pass through the building. The inner surfaces should not be covered with building materials or similar during the drying period. The straw-clay must be allowed to dry frost-free. In general a minimum of two weeks between application and possible frost should be calculated. Before the panels have received the final plaster they must be protected against heavy rain and splash water bouncing off scaffolding boards or similar.

### 10.3.3.2 New panel infill with earth brick masonry

Before beginning with earth construction works, 15 mm thick triangular battens should be firmly nailed to the inner/side flanks of the timber members of the timber frame. The triangular cross-section is not as susceptible to warping and are also easier to fit to the irregular surfaces of the beams. Larch or hardwood battens are recommended, fixed with rustproof nails. A batten on the bottom frame member is only necessary for very large panels and can be made out of a flat strip of wood with a cross-section of ca. 5 × 25 mm. The insertion of a batten on the underside of the top

 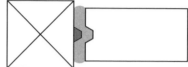

Fig. 10.62 : Diagram showing the principle of triangular and trapezoidal battens

frame member can make it difficult to complete the top course of masonry; its necessity should be considered carefully.

As an alternative to using triangular battens, channels can be cut, milled or chiselled into the beams. As mentioned above, one should not cut continuous channels along the entire length of timbers prior to assembly or inserting them into the framework to avoid the creation of moisture channels and cavities within the timber joints. Sufficiently deep existing grooves can also be used. When constructing the earth brickwork, these channels are carefully filled with mortar. These then harden to form an immovable mortar key. To aid construction (see above), these are recommended for the underside of the top frame member of the panel.

Whether battens or channels, these should lie in the centre of the thickness of the brick, i.e. taking into account the additional thickness of the external render about 6 - 7 cm from the outer face of the timber members (figure 10.62). The insertion of grooves into the ends of the bricks is only required when using rectangular or trapezoidal battens: with triangular battens the mortar is able to displace to either side of the batten when the brick is pushed up against it; with flat-faced battens, space needs to made in the brick to accommodate the mortar.

In earlier times the above method served to mechanically fix the panel of brickwork in the frame primarily during construction. The techniques used were correspondingly simple. It is also possible to insert rustproof nails into the posts approximately every 25 cm to create a sufficient connection. More technically complex solutions such as seals inserted under the battens are not found in old timber-frame panels. The grooving of bricks where they meet the posts of the panel were likewise virtually unknown. Justifications such as windproofing and protection against driving rain are more modern concerns. The following discussion examines these in greater detail:

The development of more ambitious junction details often assume a somewhat idealised notion of historical timber-frame constructions. Unlike many cleanly executed diagrams, many old timber members do *not* exhibit clear rectangular cross-sections. They are also not always equally thick. The flanks of beams often exhibit differently

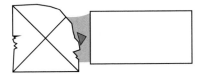

Fig. 10.63 : Diagram showing the more realistic geometry of many timber-frame members

sloping rough edges or have deep ridges and diverse indentations or protrusions (figure 10.63).

If complex detail solutions are to be realised properly, this requires elaborate individual adaptations of the bricks, battens and seals. Even when well executed, holes can still easily remain. At the corners where the battens meet, they need to be mitred to avoid problems in these particularly susceptible junctions. If this were not undertaken its function as a rainwater seal would be compromised by the ability of water to enter the end grain, endangering the head piece.

Where frame members exhibit irregular cross-sections, the best one can expect is to minimise the leakage between beam and brickwork panel, even with excellent workmanship. The work required to achieve this is considerable. Where framework panels are small and narrow, such complex details can become impossible to realise (figure 10.64). In unfavourable conditions the insertion of rubber membranes may hinder the ability of the panel to dry out any moisture that has managed to penetrate the panel.

There are therefore several reasons why the function of the triangular battens should be limited to the simple mechanical securing of the brickwork panel. The limited ability to protect against driving rain is better achieved through good workmanship in the application of the external render. Effective windproofing and airtightness is only possible through the creation of continuous surfaces on the internal face such as plaster layers.

For the durability of new infill panels, especially in areas exposed to weathering, the choice of suitable earth bricks is important. The *Lehmbau Regeln* divide earth bricks according to the requirements they should fulfil into *Usage Classes* (see chapter 8). For plastered surfaces exposed to the weather, earth bricks of *Usage Class I* should be used (figure 10.65). They must have a homogenous solid structure and adequate resistance to water and frost. Deformations resulting from swelling and shrinkage may only occur to a small degree. Extruded earth bricks often (but not always) have unfavourable properties in this respect. Bricks of *Class I* should not be perforated. If the above properties are fulfilled, a degree of perforation of up to 15 % can be tolerated. Grille blocks

Fig. 10.64 : Timber-frame building with numerous small and non-rectangular panels (Berg-Nassau, Lahntal)

with more than 15% perforations are not permissible. *Green unfired bricks*, bricks that are destined for firing in brick production, may not be used. The classification of commercially produced bricks is the responsibility of the brick manufacturers.

A further important criteria for the suitability of an earth brick for timber-frame constructions is the roughness of the external face of the bricks. Lime renders do not adhere well to bricks with very smooth surfaces. Experience has shown that even the grooved surfaces of very smooth bricks, as moulded in the surface of perforated bricks, do not provide sufficient surface roughness.

The adhesion of the external render has proven to be better for masonry made with small brick formats (≤ 2DF) than large-format bricks due to the large number of mortar joints.

Fig. 10.65 : Earth bricks of Usage Class I as mould-pressed bricks with coarse organic additives

The execution of the brickwork follows the generally accepted rules of the bricklayer's trade. The horizontal and vertical mortar joints should not be thicker than usual with the exception of the joint around the perimeter of the panel which can be ≤ 3 cm thick. For vertical mortar joints, the typical brick overlap should be observed. Similarly the position of the face of the wall of masonry with respect to the face of the timber-frame construc-

Fig. 10.66 :
Fresh earth brick masonry in a timber-frame panel with scratched out mortar joints

tion should be carefully executed to ensure that the subsequent plaster can be applied with a uniform thickness over the entire face of the panel. After completion of the masonry the still fresh mortar joints are raked to provide a better mechanical key for render adhesion. This should be undertaken with jointer to a depth of about ½ cm. V-shaped joint recesses cut with a trowel are less effective than square recesses (figure 10.66).

Where masonry is made of high-quality bricks a moderate degree of weathering between completion of the wall infill and rendering is generally not a problem provided that brickwork that has become wet is not be subject to frost. The surface should be protected against driving rain and splash water from scaffolding boarding.

## 10.3.4 External render

### 10.3.4.1 Exposed timber-frame constructions and weathering

Exposed timber-frame constructions have limited resistance to weathering. This applies first of all to external render, although damages to render cannot automatically be attributed to inappropriate execution. They may also be a sign of an overburdened façade and simply an early indication of other causes. Sustained exposure to moisture will destroy the timber-frame members. Heavily weathered façades were therefore rendered over the entire surface or protected by weatherboarding. The problem of sufficient adhesion of lime render on earth substrates should be regarded with this in mind. An impervious connection between the background and the render may seem ideal, but in principle the render adhesion must only be as good as one can reasonably expect of a façade that is executed as an exposed timber-frame construction.

A rain load of less than 140 litres of driving rain per year is generally not problematic. Measurements have shown that the average rain load in driving rain load class I according to DIN 4108 Part 3 lies beneath this value. For buildings exposed to extreme weathering or windward façades, however, this value can be exceeded. In principle, exposed timber-frame façades are also possible in regions with higher driving rain loads if they are in sheltered locations such as the densely-built centres of towns and villages [Eckermann/Veit, 1996].

A generalised classification of where exposed timber-frame façades are permitted is therefore of little practical use; the appropriateness should be decided in each individual case. The exposure to weathering should not be viewed in isolation; rather the *relationship* between weathering and the ability of the structure to dry out. In this regard, the capacity to dry out properly can be the more important criteria [Eckermann/Veit, 1996].

A degree of experience and careful observation is required to assess the degree of weathering a timber-frame façade may be exposed to. The following aspects offer useful indications:

- Location in the landscape (sheltered or exposed)
- Orientation
- Sheltered by neighbouring buildings, hedges and trees
- The condition of the panel infill and surface of the timber frame (weathered paint and render, washed out areas, soiling, moss and algae growth)
- Proportion of weathered timber-frame members that need replacing
- Condition of the timber-frame façades of buildings in the immediate vicinity
- Traces of earlier weatherboarding or full surface plaster (nail holes, axe marks, remains of plaster lath, wire mesh, absence of paint remains on timber-frame members)
- Typical surface treatment of surrounding buildings and buildings in the region e.g. whether weatherboarding or full façade render is more typical.

The treatment of façades can also have architectural or visual/aesthetic reasons, but in many cases there are also practical reasons for regional building traditions.

The following description of rendering procedures applies for all cases where the fundamental decision to leave the timber-frame construction exposed has been considered justifiable. Further measures are described under *Special case: rendering with plaster lath mesh* (section 10.3.4.2).

As a rule, timber-frame constructions should be exposed on only one side (externally). An exposed timber frame on the outside and inside should not be undertaken when the building is heated as it is not possible to achieve a windproof and sufficiently thermally-insulated construction with reasonable constructional and economic means. In addition, the exposed treatment of irregular timber-frame internal walls demonstrates a lack of understanding of historical construction methods and conflicts with building conservation considerations.

### 10.3.4.2 Rendering timber-frame panels

#### When to apply render

The external render should be applied as late as possible so that the probability of further movement in the timber-frame construction and deformations of the timber members as a product of swelling and shrinkage has subsided. Ideally one should wait until the construction has experienced an entire heating period. In individual cases, it may be necessary to implement temporary weathering protection measures for the façade.

#### Choice of render mortar

The panels of non-weathered timber-frame façades can be rendered externally with earth mortar. For weathered walls, however, earth renders are not resilient enough. The use of admixtures and coatings to improve the weathering performance of earth plasters should only be undertaken by those with appropriate experience.

Lime mortars should be soft and elastic. As a binding agent, all manner of particular kinds of lime are recommended depending on their origin or means of storage (shell lime, pit lime). The choice of lime determines the render properties to a certain degree, the kind of admixtures and the mixing proportions are equally important.

Suitable renders include those that like historical mortars are solely lime-bonded and exhibit a stable grain skeleton with a coarse sand component (grain size approx. 1/3 of the thickness of the render coat). The generous addition of animal hairs is also recommended. The compressive strength should be around 1 N/mm². Hydraulic admixtures make the mortar too hard and cement-bonded renders are not appropriate.

#### Preparation of the render substrate

The render substrate needs careful preparation. Over and above the normal adhesion requirement between background and render, façades exposed to increased weathering may have to withstand greater stresses: the moisture uptake of timber construction members causes them to swell and apply lateral pressure to the lime render pan-

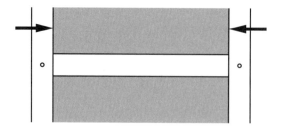

Fig. 10.67 :
The effect of lateral pressure on the panel render

el, which in serious cases can lead to outward bulging and separation of the render (figure 10.67). Narrow panels on heavily-weathered surfaces (roof verges, gable peaks) are particularly susceptible.

The roughening of the earth plaster base to provide a sufficient mechanical key has been described earlier.

As a further means of preparation it is possible to work in a slurry of lime and sand into the adequately wetted earth substrate (section 10.3.1.4). The slurry must be sufficiently binding so as not to impair the adhesive bond. The mixture can be rubbed in with a hard painter's brush or a felted, plastic or wooden float. The slurry application should be protected against too rapid absorption.

An alternative approach is the method proposed by *Gaul*: a bonding layer between the earth and lime can also be achieved by applying several applications of lime-casein paint (1 part low-fat quark, 1 part lime, 8 - 10 parts water). A cement spatter coat (1 part cement, 3 parts sand 0/4) can be applied in a network with a maximum coverage of 50 % [Gaul, 2000].

The water retentivity of pure lime mortar, where necessary, should not be adjusted using admixtures as with other factory-produced ready-mix mortars. If the plaster base draws moisture out of the lime mortar too quickly, the underlying layers do not cure properly. The plaster base must be sufficiently and carefully pre-wetted (spray mist application) panel for panel, directly before application of the render coat. Too much pre-wetting causes water to saturate the pores of the substrate, hindering the process of mortar adhesion.

### Render build-up, application and surface finishing

For surfaces exposed to normal weathering, two-coat plasters with a total thickness of 15 mm are suitable. In more sheltered areas, a thin coat of up to 5 mm, as seen on historical buildings, can be sufficient. Well roughened straw-clay or firm, sufficiently-rough earth brick masonry serves as a better render substrate than earth render. Render build-ups with a base coat of lean earth plaster mortar and a topcoat of lime mortar are correspondingly unsuitable.

The application of the lime render should follow the instructions given by the manufacturer and the general rules of the trade. When hand-processed, the mortar can be spread onto the panel with a wood plasterer's float (15 × 40 cm) and worked into the surface with a firm side-to-side movement of the float. An alternative is to throw on the mortar with a masonry trowel. The surface of the lime base coat must be sufficiently rough for the subsequent application of the topcoat render (figures 10.68 and 10.69). The topcoat can be applied with a finishing trowel after the first coat has cured.

The lime mortar should be protected against too rapid water evaporation by wind and sun. Methods include arranging hangings over the surface and gentle re-wetting. Note that the fresh lime plaster may not be covered with a continuous film of moisture as the water saturation hinders the diffusion of carbon dioxide. For this reason, the surface must be protected against longer exposure to rain.

### Panel seams

The render should be extended as far as the edges of the panels, ideally in the full thickness. The aim is to achieve a render surface that lies flush with the front face of the timber-frame construction. Where timber-frame panels bulge outwards the render should not grow too thin towards the edges. The render should meet the chamfered edges or unfinished surfaces of beams at a 90°-angle. The edges of the beams should lie just 1-2 mm forward of the plaster (a good edging for subsequent painting).

The fresh render can be separated from the wood by a fine, approx. 2 mm deep knife or trowel cut. This prevents uncontrolled cracking that can arise when the lime render sticks to the forward edge of the wood. In many historical timber-frame buildings

Fig. 10.68 : Lime mortar applied with a wood plasterer's float or thrown with a masonry trowel

Fig. 10.69 (left): Good mechanical key between earth bricks and lime mortar

Fig. 10.70 (right): Damage to timber members where mastic has been applied to the panel seam

deeper trowel cuts are not practically possible due to the irregularity of the side flanks of the timber-frame posts and beams. The separation should be cut cleanly and not damage or decimate the render along its edge. Chamfered edges of plaster sections are not suitable as they channel water directly into the panel seam. The filling of the junction with mastic should be categorically rejected. It does not provide long-term protection against moisture ingress and is more likely to hinder the ability of the panel to dry out properly at a critical point (figure 10.70).

### Render coatings

To reduce the frost susceptibility of lime mortars, they should be given a protective coating. The coating material or paint should be compatible with the soft alkalinity of the render material. Likewise, the coating should have a degree of vapour permeability as film-forming paints would direct the entire rainfall into the panel seam at the bottom. Paints and coatings with a water vapour resistance of $S_d < 0.1$ m and a water absorption coefficient, w of between 0.3 and 1.0 kg/m²h$^{0.5}$ are recommended [WTA, 1999].

### Special case: rendering with plaster lath mesh

As described previously, very heavily weathered surfaces should not be realised as exposed timber-frame constructions to ensure the long-term structural integrity of the timber members. In practice, however, architects and craftsmen are regularly asked to realise such façades or are confronted with borderline cases.

The use of dense plaster meshes such as expanded metal mesh on earth plaster surfaces can lead to damages as their dense mesh cannot be penetrated by coarse lime mortars. Renders and sheet-like inserts form rigid membranes that run the risk of separating from the background allowing water to run behind them. The panels and timber-frame construction may then be subject to prolonged moisture damage that goes undetected for years.

For such purposes a large-mesh plaster lath made of stainless steel with regular indentations is suitable (figure 10.71), fixed with long rust-free screws into the panel. The mesh should not be fixed to the beams as movement in the timber construction can lead to damages.

Despite the use of a plaster lath mesh, the render substrate still needs careful preparation as described above. Similarly, the render must be protected against too rapid dehydration. A two-coat render should be used.

Plaster reinforcement meshes made of coated fibreglass do not support the render but reinforce it. They therefore do not help improve the adhesion of the lime render to the earth substrate but they can be used to resist crack formation.

### 10.3.4.3 Rendering and cladding entire façades

Timber-frame façades can be *rendered over their entire surface* or *clad with a hung façade* such as weatherboarding.

After completing any necessary timber repair works, the panel infill should be repaired with materials appropriate to the existing building fabric and holes and irregularities should be filled. The members of the timber frame and the panels must have attained their equilibrium moisture level before applying a render coat.

The mechanical separation of timber surfaces from the render coat is a common requirement. Typically the surfaces of timber-frame members are covered to avoid movement and deformations in the timber frame from causing cracks in the render. Vapour-permeable papers or textiles should be used, vapour barriers such as bituminous paper or oil impregnated paper are unsuitable [WTA, 2000]. In many cases, however, the mechanical separation is unnecessary, for example

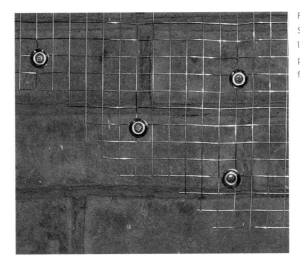

Fig. 10.71 : Stainless steel plaster lath mesh (special product for timber-frame repairs)

in façades that are largely intact and where the structure has not been changed and where fibre-reinforced render mortars are used.

All non-rusting materials that have a wide enough mesh to allow full capillary contact between the plaster and substrate are suitable for use as a plaster lath. Rolled reed plaster lath with approx. 70 stems per m have proven themselves over years of use. The plaster lath should not be fixed to the beams but wherever possible to the panels. Long rustproof screws are suitable for fixing with the kind and density of fixtures depending on the solidity of the timber-frame panels. The plaster lath should sit firmly and not yield elastically.

For the render, vapour permeable materials such as lime mortar are suitable. Thermal insulation renders can also be used. Lime mortars can be reinforced with animal fibres. The render should be applied and treated according to the manufacturer's instructions. The incorporation of a suitable reinforcement mesh for weathered lime plasters in the upper third of the render build-up is strongly recommended as long as the plaster base does not function as tension reinforcement in this zone. The water vapour resistance of the external render should be $S_d = 0.3$ m, that of the external paint no more than 0.1 m [WTA, 2000].

Systems consisting of *render on insulation boards* should be vapour permeable, including any adhesive or levelling compounds used. When fixing insulation boards, care should be taken to observe the limited mechanical loadbearing capacity of the earth timber-frame panels.

*Hung cladding* includes wall coverings made of timber boarding or slate tiles. Any insulation behind the cladding should be vapour permeable. To reduce the eaves and verge overhang, cladding that has a large proportion of open joins sometimes does not exhibit typically dimensioned rear ventilation. In such cases it is important that condensation water cannot form either within the insulation or on the internal face of the hung cladding.

## 10.3.5 Building material and building element properties

### 10.3.5.1 Mechanical properties

All of the panel infill techniques described above function as a spatial enclosure, weather protection and thermal and sound insulation. They do not serve a stiffening function for the structural framework.

Fixtures in old earth infill panels should be made with sufficiently long and thick screws (L ≥ 120 mm, Ø ≥ 6 mm). Nails and anchors are not suitable.

## 10.3.5.2 Thermal insulation and water vapour diffusion resistance

Timber-frame walls are elements that consist of several combined components. The individual components must therefore be considered separately in terms of their thermal insulation. The following descriptions therefore differentiate between
- *Panel infill core* made of wattle or staves with straw-clay daub or earth brick masonry (table 10.6, 10.7)
- *Panel facing* made of straw-clay and render (table 10.8)
- *Timber-frame post/beam* (table 10.9)

The *bulk density* of historical timber-frame panelling of wattle or staves with straw-clay, which is relevant for their thermal properties, can only be estimated; an exact analysis would be elaborate and costly to ascertain. The estimation appears to be permissible as the differences have only a moderate effect on the overall thermal conductivity of the wall. The *bulk density* of earth brick masonry can and should be ascertained.

Table 10.6 : $1/\lambda$-values for an 8 cm thick panel infill core made of wattle and straw-clay daub
($\lambda$-values for earth building materials and wood according to DIN 4108-4 2002-02) [DIN 4108, 2002]

| Bulk density, straw-clay<br>kg/m³ | Wattle and straw-clay daub<br>D = 8 cm (empirical value)<br>m²K/W | Staves with straw-clay<br>D = 8 cm (empirical value)<br>m²K/W |
|---|---|---|
| 1100 | 0.21 - 0.23 | 0.28 |
| 1300 | 0.17 - 0.18 | 0.22 |
| 1500 | 0.14 - 0.15 | 0.18 |

Table 10.7 : $1/\lambda$-values for a 12 cm thick panel infill core made of earth block masonry
($\lambda$-values for earth building materials according to DIN 4108-4 2002-02) [DIN 4108, 2002]

| Bulk density, earth bricks<br>kg/m³ | Earth brick masonry<br>D = 12 cm (empirical value),<br>m²K/W |
|---|---|
| 1400 | 0.20 |
| 1600 | 0.16 |
| 1800 | 0.13 |

Table 10.8 : λ-values for panel facings of straw-clay and plaster
(λ-values for earth building materials and lime plaster acc. to DIN 4108-4 2002-02) [DIN 4108, 2002]

| Bulk density, material kg / m³ | λ W / m²K |
|---|---|
| 1100, straw-clay | 0.41 |
| 1300, straw-clay | 0.53 |
| 1400, straw-clay | 0.59 |
| 1500, straw-clay/earth plaster | 0.66 |
| 1600, straw-clay/earth plaster | 0.73 |
| 1800, lime plaster | 0.87 |

Table 10.9 : λ-values for timber-frame members (λ-values of wood: DIN 4108-4 2002-02) [DIN 4108, 2002]

| Bulk density, kind of wood | Timber-frame post/beam |
|---|---|
| 500, softwood spruce/fir, pine | 0.13 |
| 700, hardwood oak | 0.18 |

Water vapour diffusion resistance

DIN 4108 specifies a general water vapour diffusion resistance coefficient for earth building materials of 5/10 [DIN 4108, 2002].

10.3.5.3 Sound insulation

According to calculations by the SWA-Instituts in Aachen, a value of 41 dB can be estimated for the assessment of the sound insulation [Breidenbach/Röhlen, 2008]. The above estimation is based on the following construction:
- Timber frame: 14 cm (ratio of panel surface:beam surface = 70 %:30 %)
- Beams: oak/beech, external render: lime 2 cm
- Internal plaster: earth, 2 cm

10.3.5.4 Fire performance

Building material class

Earth bricks and mortar *without* organic additives are classified according to DIN 4102 Part 4 as the equivalent of non-flammable [DIN 4102, 1994].

If the earth mixture contains organic additives or the panel infill core contains wooden components, they are not classified in DIN 4102-4. Further details on the flammability of straw-clay mixtures are contained in the *Lehmbau Regeln* [DVL, 2009]: straw-clay mixtures with a bulk density > 1200 kg/m³ have been demonstrated in normed tests to correspond to the equivalent of *non-flammable*, while mixtures with a bulk density > 600 kg/m³ correspond to *flame-resistant* materials.

Fire-resistance rating

DIN 4102 Part 4 classifies timber-frame constructions under particular conditions as F 30 [DIN 4102, 1994]:
- Timber-frame beams ≥ 100 × 100 mm when exposed to fire from one side, Timber-frame beams ≥ 120 × 120 mm when exposed to fire from two sides
- Timber-frame panels fully enclosed with an earth covering
- Cladding on at least one side. Permissible claddings include gypsum fibreboard panels ≥ 12.5 mm, gypsum plasterboard ≥ 18 mm, wood-wool insulation board ≥ 25 mm, wood-based boards ≥ 16 mm or tongue and groove boarding > 22 mm.

Also permissible is a coating with ≥ 15 mm thick plaster/render in accordance with DIN 18550 (valid at the time of introduction DIN 4102 Part 4). Earth renders or straw-clay coatings were not described in DIN 18550; see chapter 3 *Earth building materials* for further details of their properties.

The WTA technical information sheet *Merkblatt 8-12-04/D Brandschutz bei Fachwerkgebäuden (The fire protection of timber-frame buildings)* cites fire tests that propose the classification of timber-frame walls with external render and internal plaster (i.e. covered on both sides) as F 60 [WTA, 2004].

## 10.4 Timber beam ceiling infill

Whether in historical timber-frame buildings or monolithic constructions, for hundreds of years there was practically no alternative to timber beam ceilings. Only in the 19th century did steel girder constructions become available and still later reinforced concrete. But well into the middle of the 20th century, ceilings with inserted boards and earth filling were commonly used. The boards and fill material served as an enclosure for the room, as cavity filling for thermal insulation and as sound insulation and fire protection.

As with earth panel infill of timber-frame constructions, ceiling constructions are an important component of the authentic fabric of historical buildings. In addition, the articulation of the underside of the historical ceilings creates an impression on the room below.

### 10.4.1 Traditional ceiling infill techniques

The ceiling infill techniques described below – *staves with straw-clay infill* and the variant with *earth reel infill* – can be found in many historical timber-frame buildings as well as in monolithic earth constructions. The space between the rafters should likewise be filled in a similar manner. *Ceiling inserts with loose earth infill* are more typical for historical monolithic constructions from the 19th century onwards.

#### 10.4.1.1 Staves with straw-clay infill

To hold the staves, *grooves* are cut into the side flanks of the ceiling beams. The position of the groove on the vertical faces of the ceiling beam determines the position of the staves or other inserts in the cross-section of the ceiling and with it the profile of the underside of the ceiling. If the groove is arranged toward the top of the beam, almost the entire height of the beam is visible from beneath. If the groove is positioned further down, the position of the beams may only be visible as a strip of plaster on the underside of the ceiling. A very common arrangement is to locate the groove between half-height and the upper third of the beam cross-section. In the low mountains of the Rhine, for example, the height of the groove is ~3 cm, the depth ~2 cm (figure 10.72).

Fig. 10.72 : The location of the groove in the cross-section of the beam determines the profile of the ceiling construction and with it the articulation of the underside of the ceiling

The material and manufacture of the *staves* follows the same principle as used for straw-clay daub in timber-frame construction (section 10.3.1.1). The staves are split with an axe, worked and then cut to the right length with pointed ends. Where the ceiling beams were parallel, a whole series of staves could be made in one go. Their length had to be just enough to sit tightly between the grooves. To prevent the smoothed plaster on the underside from being too thick, the staves were made as flat as possible on their underside, either by careful selection, appropriate working or the cutting of the pointed end. The staves were spaced a few centimetres apart, wide enough to push the straw-clay mixture between them but close enough to avoid the straw-clay from sagging. To wedge in the last one to two staves, which could not be inserted diagonally, a notch was cut into one side from above so that the stave could be pressed down into the groove and brought into position (figure 10.73).

Fig. 10.73 : To insert the last one or two staves into a panel between two beams, a notch was cut from above on one side, making it possible to insert the staves almost perpendicularly

Alongside a closely spaced lath of staves, wattles were also sometimes used in an arrangement similar to that used for timber-frame panel infill (section 10.3). Later, from the 19th century onwards, a variety of different methods arose, such as the laying of staves on battens nailed to the side flanks of the beams. In some cases continuous battens were laid across the tops of the ceiling beams, often in the form of split logs or the halved trunks of straight, young coniferous trees. Evidence of continuous battens nailed to the underside of the timber beams have also been found.

The panels between the beams were usually filled flush with the top surface of the ceiling beams with straw-clay mass (figure 10.74), but sometimes the infill stopped short of the top of the beam with a layer of air between.

The straw-clay was filled between the beams in a malleable, viscous consistency and pressed between the staves. On the underside, the loose strands of fresh straw-clay hanging between the staves were then pressed back onto the underside of the staves until they were completely covered. The adhesion of the straw-clay to the underside of the wood surface was not always ideal and one can sometimes find twigs nailed to the underside as a plaster lath. Sometimes straw-clay layers on the underside appear to adhere only because they are held at their edges by the side flanks of the beams.

Fig. 10.74 : Underside of staves with straw-clay infill: the underside of the ceiling of this farm building did not need plastering and the straw-clay has not been pushed through

### 10.4.1.2 Earth reels

In some regions, staves wound with straw were also called *weller*. To avoid confusion with monolithic weller earth construction, the term used here is a *Wickelstake*, or *earth reel*.

The staves are usually wedged between the beams in grooves as described earlier in *staves with straw-clay infill*. Because the staves cannot be trimmed or worked after they have been coated, the assumption is that the staves were first cut to size and inserted between the ceiling beams to test their fit, then removed again for coating with straw-clay.

In later examples, the staves are not wedged into grooves but laid on battens that are nailed to the sides of the timber ceiling beams.

The trimmed staves were wound with long-stem straw laid in a soft, viscous mass of prepared earth. Alongside different regional preferences, the kind of straw, the harvesting length or the emergence of new agricultural machinery (for example thresh-

Fig. 10.75 : Different ways of making earth reels

Fig. 10.76 :
Underside of an earth reel ceiling that has been opened up. The roughening of the timber beams with an axe has improved the plaster adhesion but only with limited success.

ing machines) may have led to the development of different winding techniques (figure 10.75). Before winding, the straw was left for several hours or overnight for it to become soft and supple. The formable straw was then laid out as a mat and then wound onto the stave. In another method, the lengths of straw were wound around the stave in ring-like segments.

A further common method was to form 'ropes' and then wind these 1 ½ times in a helical screw movement around the stave, leaving 15 to 20 cm hanging down at the end. After mounting the staves, the ends could be pressed smooth on the underside. Another option was to wind the straw 'rope' entirely around the stave (figure 10.75).

The diameter of the *wound earth reels* varied between 8 and 15 cm. For large crosssections the staves were sometimes enclosed in a malleable mixture of earth and chopped straw and then additionally wrapped with a long-stem straw mat.

The staves were brought into position in the ceiling while still wet and pressed against each other, sometimes with so much pressure that the round cross-section of the moist-malleable reel became pressed into an oval shape.

While still moist, or after drying, the underside of the surface was smeared with strawclay and smoothed over to form a flat underside (figure 10.76).

Fig. 10.78 : Underside of an opened-up ceiling insert

### 10.4.1.3 Ceiling insert with loose or compacted earth infill

An alternative method of filling ceilings was to install ceiling inserts made of boards between the ceiling beams. They rested either on a rebate or on battens nailed to the side flanks of the ceiling beams (figure 10.77).

In older more prestigious buildings, the visible underside of ceiling inserts made of oak or other boards have been profiled or decorated with coloured frames. The inserts between the timber ceiling beams of buildings in the late 19th century consisted mostly of simple softwood boards or planking. When the underside was to be plastered it was more common to create a flat underside to the ceiling rather than to plaster around the ceiling beams. The smooth underside was also a design aspect as the flat surface provided a free 'canvas' for stucco decoration. If the boards were located lower down the beam cross-section, they could be used as a plaster lath. To improve plaster adhesion the surfaces were worked with an axe. Sometimes thin branches were nailed to the beams under which a plaster lath might be jammed as in the example in fig-

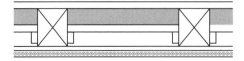

Fig. 10.77 : Cross-section through a ceiling with ceiling inserts

Fig. 10.79 : Section through a ceiling in a dilapidated timber-frame house (Heuersdorf, Saxony)

ure 10.78. A thick layer of earth plaster was applied to this simple plaster lath onto which several thin layers of lime were applied.

Where the boarding of the insert and the underside of the beams did not coincide, a separate supporting construction needed to be created for the plaster. Typically thin lattices of *timber laths* were used, or alternatively planks were nailed to the underside of the beams and then covered with *reed plaster lath*. The flat plaster over the entire underside also served to prevent particles from the ceiling fill from falling into the room below.

The ceiling inserts carry the weight of the ceiling infill. Before the fill material was inserted, the top surface was often smeared with a layer of malleable earth or straw-clay of about 2 cm thick.

In the image above one can see a 1-2 cm thick layer of chaff above the ceiling boards which alternate with split logs to form the underside of the ceiling insert (figure 10.79). The supposition is that the chaff was strewn dry into the insert to protect the wood against moisture from the wet mass of the straw-clay that followed.

The earth used for the fill material was typically in a dry or naturally-moist consistency and could be poured over the insert and compacted. From the 19th century onwards, loose fill material in the form of dry construction waste or slag was also used. The floorboards of the floor surface above were typically nailed directly to the top surface of the ceiling beams.

### 10.4.2 Repair of old ceiling infill

#### 10.4.2.1 Repair of ceiling infill made of staves and straw-clay or earth reels

If repairs to ceiling infill made of *staves and straw-clay* necessitate the replacement of staves, these must be individually fitted and inserted. For shorter sections or individual replacements, the ceiling needs to be opened and material removed from above. The same procedure applies for replacing *earth reels*. Where necessary, replaced sections of staves may require the application of a plaster lath on the underside to receive the subsequent layers of earth ceiling plaster. Replacement of the staves with panels or large-format boarding is not in keeping with the structure but technically possible.

#### 10.4.2.2 Repair of timber boarding with loose or compacted fill material

Damaged or missing boards should be replaced. If the infill is to be inserted in a wet or moist consistency, the boards should be spaced slightly apart (< 5 mm) so that they do not buckle when they swell. The replaced sections should be covered to prevent dusting. Replaced infill material should correspond as far as possible to the existing materials.

### 10.4.3 New ceiling infill

#### 10.4.3.1 New ceiling infill made of staves and straw-clay or earth reels

New ceiling beams can be grooved in the workshop before installation. Oak or softwood battens (for example roof battens) are usually used today for the staves. Staves recovered from the existing construction can also be reused. Before insertion they are pointed with an axe or suitable saw. Around 10 - 12 staves are required per linear metre ceiling panel.

The staves are brought into place with a heavy hammer or the reverse side of an axe and spaced apart by about 2 - 4 cm. The staves should fit tightly but not force apart the ceiling beams. The last stave is fitted via an extra notch cut from above.

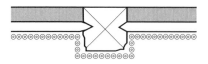

Fig. 10.80 : Reed plaster lath can be fixed to the ceiling panels and beams and, if desired, the irregularity of the ceilings beams can be 'corrected'

The old straw-clay infill was pressed through the spaces between staves and smeared flat from below. However, as the mixture does not adhere well to the smooth sawn surfaces of the underside of the staves, the use of a coarse-mesh plaster lath is recommended, for example wire-bound reed plaster lath. The lath can be fixed to the entire underside of the ceiling, and any irregularities in the beams, if desired, made more uniform (figure 10.80). The plaster lath simultaneously serves as permanent formwork for the spaces between the staves and provides a good plaster lath for the subsequent ceiling plaster. When using reed plaster lath it is important that the thicker carrier wire is stapled to the staves not the thinner binding wire, i.e. the mat should be mounted with the thicker carrier wire on the underside (see section 4.3.4).

The straw-clay for the fill material can be manufactured on site or sourced ready-to-use from a manufacture. Aside from conservation requirements that new materials correspond to existing materials, there are few specific requirements for the infill material. The straw-clay mixture is typically inserted in a malleable consistency and should enclose all wooden parts of the construction and be pressed in to fill all voids and cavities. Typically the ceiling is filled until just below the top surface of the beams. For particularly thick layers, the loadbearing capacity of the beams must be verified before insertion of the material. The straw-clay must be allowed to dry out rapidly, if necessary with additional forced drying.

The manufacture of wound *earth reels* typically follows the historical method used for the existing building. Different methods of making earth reels are described above. Some manufacturers also provide pre-made earth reels. The earth reels are installed directly after manufacture, at the latest as they begin to dry. The reels can be inserted by hand but should not be under stress. The earth material at the top and bottom can be pushed slightly to one side to facilitate knocking it into place with a hammer or axe. Once in place, the earth material is slid back.

### 10.4.3.2 New ceiling inserts with earth mass or earth loose fill material

For the supporting surfaces for the boarding, sufficiently wide battens, for example 4/6 cm, are nailed with the long side parallel to the sides of the beams. Boards or planking that will receive wet or moist earth material should not be laid too tightly to prevent them from buckling as they swell under moisture. The boarding should be covered to prevent dusting, in particular at the edges and junctions where it may be necessary to fold and stick down the building paper. Reinforced building papers should be used for this purpose to avoid ripping, especially when inserting wet or moist infill material as normal paper can become brittle once dry. Plastic foils or oil-impregnated papers are less favourable as they can lead to condensation water forming on the surface of the wooden boards and mean that wet fill material can only dry from above.

Any earth mixture of a free-flowing consistency can be used as a fill material, regardless of quality or constituents. The weight of the fill material can be modified by the use of lightweight aggregates.

Before further work is undertaken on the surface covering of the floor, the fill material should be properly dry. This can be tested on site but precise results are only possible by comparing the weight of a sample before and after oven drying. To keep construction times down and reduce the amount of moisture introduced into the construction, dry fill material provided by earth building material manufacturers can be used. The exposure of the room above to dust must be limited. Earth bricks can also be loosely laid as a fill material. Any remaining gaps can be filled with loose fill material and the seams brushed full with sand.

### 10.4.4 New ceiling plaster

#### 10.4.4.1 New ceiling plaster on ceilings with staves & straw-clay or earth reels

In many regions, the entire underside of the ceiling, that is the beams and the panels between them, were plastered with earth and lime. The many different forms and means of articulation of the plaster reveal the importance accorded to the ceiling. Irregularities in the shape of beams were corrected resulting in straight, sharp-edged parallel lines. The current trend of removing ceiling plaster to create a rustic-looking interior does not, therefore, correspond to the intentions of the original inhabitants and usually looks out of place. The same applies to ceiling panelling or suspended ceilings. They change the original proportions of the room and affect the appearance of the windows in the wall surface sometimes considerably.

The easy formability of earth mortar makes it possible to shape the material to create different kinds of junction articulations and ceiling panel framing, although these may have actually originated as short-cuts to disguise poor workmanship (figure 10.81).

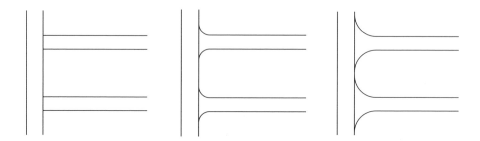

Fig. 10.81 : Rectangular beam junctions (left), "Kölner Decke" (Cologne ceiling) with quarter-circle transitions (centre), semi-circular ceiling panel ends (right)

Fig. 10.82:
Forming the quarter-circle roundings of a "Kölner Decke"

Over time these developed into decorative forms of their own that have characterised the interiors of entire regions for centuries. An example is the *Kölner Decke* (Cologne ceiling) which is to be found throughout the Rhineland region. Its name testifies to a desire to emulate the cultivated salons of urban society (figure 10.82).

Historical earth plasters and their further treatment is described in section 10.6.

Alongside painting with limewash or textured lime slurries, the underside of ceilings can be plastered with a thin skim coat of fine lime mortar. As a rule, light-coloured earth plasters are a better fit for earth undercoat plasters as the thin layers of lime plaster adhere less well and are more susceptible to scaling off. When plastering with lime, it can be beneficial to work in a well-binding lime slurry into the carefully pre-wetted earth undercoat. The fine lime plaster is applied in a thickness of ca. 3-5 mm. Soft air-hardening or slaked lime mortars are suitable (figures 10.83 and 10.84).

Fig. 10.83:
Underside of a ceiling with rectangular junctions and plaster of uniform thickness

Fig. 10.84:
Underside of a "Kölner Decke" with quarter-circle transitions and 2/3 of the beam height showing

#### 10.4.4.2 New ceiling plaster on ceiling inserts with earth fill material

For the plastering of the underside of ceilings with inserted boarding and beams, the ceiling must first be covered with a plaster lath. Rolls of reed plaster lath should be fixed with the thicker carrier wire away from the wood, i. e. on the side of the room and the workman (see chapter 4 and section 10.4.3.1).

If the inserted ceiling boards are indented, the flat underside of the ceiling panels are often clad with drywall boards or plaster base boards such as clay plasterboard or wood-wool lightweight building boards. They are usually fixed to an intermediary subconstruction, but sometimes directly on the ceiling beams (chapters 4 and 6).

### 10.4.5 Building material and building element properties

#### 10.4.5.1 Mechanical properties

The historical ceilings described above do not fulfil the requirements of stiffening elements according to current standards. However, the large number of clamping connections, for example the tight arrangement of *staves with straw-clay*, can in practice have a certain stiffening effect. Nailed floorboards have a similar effect.

When lighter fixtures are to be fixed in the fill material of ceilings, sufficiently long and thick screws should be used. They can be fixed firmly in the staves, and even in hardened straw-clay mass. In inserted ceilings, fixtures can be screwed to the boards or planks. In most cases anchors or plugs are not required.

#### 10.4.5.2 Thermal insulation

The thermal properties of existing timber beam ceilings with an earth filling are only relevant when the ceiling separates heated rooms from unheated spaces, for example, the ceiling between the top floor and the roof space. Because the ceiling is not an external construction element (i. e. does not separate outdoors from indoors), it is sufficient to approximately estimate its thermal properties.

Analogue to the system used earlier to described the properties of timber-frame walls (section 10.3.5.2), the following notes distinguish between the different components of the ceiling construction as follows:
- *Core fill material* such as staves and straw-clay or rolled earth reels (tables 10.10 and 10.11)
- *Inserted fill material* such as straw-clay or loose earth fill material (table 10.12)

The bulk densities of the ceiling fill material and the earth building materials can be ascertained by conducting tests on the materials or by simple estimation.

Table 10.10 : 1/λ-values for a 7 cm thick ceiling core infill made of staves and straw-clay
(λ-values for earth building materials and wood according to DIN 4108-4 2002-02) [DIN 4108, 2002]

| Straw-clay bulk density kg/m³ | Ceiling core infill, staves and straw-clay staves, softwood or hardwood D = 7 cm (empirical value) m²K/W |
|---|---|
| 1100 | 0.21 |
| 1300 | 0.16 |
| 1500 | 0.13 |

Table 10.11 : 1/λ-values for ceiling core infill made of rolled earth reels with a bulk density of 700 kg/m³
(λ-values for earth building materials and wood according to DIN 4108-4 2002-02) [DIN 4108, 2002]

| Diameter of rolled earth reel cm | Ceiling core infill, rolled earth reel, staves, softwood or hardwood bulk density, approx. 900 kg/m³ (empirical value) m²K/W |
|---|---|
| 8 | 0.25 |
| 12 | 0.34 |
| 16 | 0.44 |
| 20 | 0.54 |

Table 10.12 : λ-values for ceiling infills made of straw-clay and loose earth fill material
(λ-values for earth building materials according to DIN 4108-4 2002-02) [DIN 4108, 2002]

| Bulk density, material kg/m³ | W/m²K |
|---|---|
| 1200, straw-clay | 0.47 |
| 1400, straw-clay | 0.59 |
| 1600, straw-clay | 0.73 |
| 1700, straw-clay | 0.81 |
| 1800, loose earth fill material | 0.91 |

### 10.4.5.3 Sound insulation

For the approximate assessment of the sound insulation of timber ceilings, below the sound reduction index and impact sound level for two ceiling constructions according to calculations undertaken by the SWA-Institut in Aachen.

Table 10.13 : Sound reduction index and impact sound level for timber ceilings with different earth fill materials [Claytec, 2010]

| Fill material | Sound reduction index $R'_w$ dB | Impact sound level $L'_{n,w}$ dB |
|---|---|---|
| Staves with straw-clay [1] | approx. 45 | approx. 72 |
| Insert with loose fill material [2] | > 54 | < 60 |

[1] Floorboards 3 cm, Straw-clay incl. staves 8 cm, earth plaster applied to ceiling beams and panels 2 cm.
[2] Loose-fill material with a density in excess of 1200 kg/m². Ceiling structure from top to bottom: floorboards 3 cm, cavity 1 cm, ceiling fill material 12 cm, cavity 8 cm, CLAYTEC clay plasterboard 2.5 cm suspended (spring-suspenders), earth fine-finish plaster 0.3 cm; without CLAYTEC clay plasterboard underside, the sound reduction index $R'_w$ > 47 dB.

### 10.4.5.4 Fire performance

Building material class

For further information on the building material classification of historical earth building materials, see the corresponding section of the chapter on timber-frame panel infill (section 10.3.5.4).

Fire-resistance rating

In the *Lehmbau Regeln*, table T 5-10 summarises the following information regarding the fire-resistance rating of timber beam ceilings with earth fill material [Lehmbau Regeln, 2009].

*F 30 B to F 60 B:*
- *Timber beam ceiling with fully exposed beams exposed to fire on three sides*
  Ceiling overlay, for example of earth building materials of arbitrary thickness, depending on beam centres and cross-section, panelling and floor construction (cf. DIN 4102-4, March 1994 5.3.2 & table 62 for the respective individual conditions).
- *Timber beam ceiling with covered beams*
  Ceiling insert with earth covering ≥ 60 mm or staves wrapped with earth, depending on beam centres, with upper boarding and cladding beneath (cf. DIN 4102-4, March 1994 5.3.3 & table 56 and 63 for the respective individual conditions).

*F 30:*
- *Ceiling overlays* (only when exposed to fire from on top),
  Overlays of ≥ 50 mm earth mass
  (cf. DIN 4102 Sheet 4 February 1970 4.2.).

## 10.5 Earth floors

Earth floors are mainly found in buildings for livestock, barns, workshops and sheds. For threshing floors they were rationally chosen and diversely realised because the soft and elastic earth floor facilitated the hard threshing work. In houses, however, they were replaced, depending on the affluence of the residents, by wooden or tiled flooring.

### 10.5.1 Historical earth floors

In their simplest form, earth floors consist of trodden or compacted earth material. Floors with specific layering principles were also constructed. Earth floors are mentioned in older sources and in the period after 1945 they are described explicitly.

For the *dry manufacture* of earth screeds, a naturally-moist layer of rich earth is spread over the surface with a thickness of 7-9 cm. Each layer is then compacted with a rammer or tamper until no impression is left in the surface. If a lean earth material is used, the surface was soaked with *ox blood* or *tar gall*, an asphalt-like distillation product of wood tar. After a few days the surface was rammed again.

For the *wet manufacture* of earth floors, one began first with a capillary break layer of coarse gravel that was lightly compacted. A layer of about 12 cm thick rich clay was then rammed on top of this. Shrinkage cracks were closed by compaction during the drying process. The surface was then coated with a rich clay slurry. The finished surface was a coat of fine clay and ox blood or blood from another animal, horse urine and *hammer scale*, a fine iron oxide by-product of forging [Breymann, 1881].

The DIN 1965 from the year 1926 describes earth screeds with a thickness of about 15 cm for barn threshing floors, sports halls and similar buildings. To ensure that the completed threshing floor remained moist and elastic, sea salt was mixed into the earth used for the top surface. To form a hard surface, the ramming in of hammer scale was proposed. As an alternative to using sea salt in the uppermost earth mixture, a coat of ox blood or tar gall was permitted which was then strewn with hammer scale and compacted into a crust [DIN, 1965].

For earth barn floors subject to heavy loads, earth floors with a thickness of 30-35 cm was recommended. A moisture barrier layer of rammed earth was followed by a hardcore layer, a capillary break and three rammed layers of earth. The last layer of earth was drawn off level and compacted. These floors exhibited a high degree of elasticity and lasted for a long time [Polack/Richter, 1952].

## 10.5.2 The repair of historical earth floors

There is little experience of what constitutes the correct repair of historical earth floors. One can only follow, and where necessary modify, the work steps described above as best possible. When augmenting old earth floor surfaces with new material, it is practically impossible to avoid differences in the texture of the earth surfaces. For hardening and stabilising the surface, earth slurries can be used as well as *hard oil* or *half oil floor finishes*.

## 10.6 Historical earth plasters

### 10.6.1 Description of historical earth plaster methods

Internal plasters made of earth were used widely throughout Central Europe. They can be found in regions in which fired brick or natural stone were used for the building of external walls.

Through the application of the malleable mortar to the walls and ceilings, flat and smooth surfaces could be created which enhance the quality of the space. The plaster also helped to improve the comfort of the space: the surface of earth plasters with a straw content are warmer than higher-density natural stone surfaces. Internal earth plaster also helped improve the thermal insulation, windproofing and fire performance of historical timber-frame buildings. Earth was also used for the external render of timber-frame panels (section 10.3.1.1).

#### Mortar composition

*Earth plaster mortars* differ from *straw-clay* in that they contain a lower straw content and the chopped straw stems are much shorter. In most cases they were also worked into a more homogenous mass. Earth plasters for topcoat plasters could contain fine plant fibres such as flax fibres, with which it was possible to produce remarkably smooth and, even by today's standards, high-quality fine finishes. Historical earth plaster mortars without any fibrous additives are not known. Animal hairs are likewise not commonly found – they were usually added to lime renders. Sand was also used to lean down mixtures.

In the period after 1945, various authors document methods of making earth plaster mortars that reflect currently available experience of historical plasters. For the earth undercoat a richer earth mixture was used with coarse sand (no fine sand) and 2 - 3 cm long strands of coarse-cut hay or straw. The topcoat plaster was manufactured in the same way but with finer fibrous materials such as chaff or flax fibres [Niemeyer, 1946]. Others recommended beginning by preparing earth slurries. The earth should be sieved or the slurry passed through a 10 mm sieve. To this slurry, barley chaff and waste material from flax processing was added, thereafter sand [Pollack/Richter, 1952].

#### Plaster background, layering and processing

Traces of specific preparations of mineral plaster substrates for receiving earth plasters have not been found. The adhesion of the plaster to timber beams was assisted by working the surface with an axe so that scales protruded that act like a plaster lath (figure 10.85). The effectiveness of these measures was fairly limited and timber posts and beams are usually the weak points of historical earth plasters. No special measures were required to adhere earth surfaces to each other.

The plaster could be applied in layers of between several centimetres and just a few millimetres. In many cases, ever new layers of plaster were added to older layers so that thick packets of plaster resulted. Due to its plasticity and good adhesion, earth mortars were also used for thick applications and working overhead on ceilings. Numerous complex stucco ceilings have an undercoat of earth plaster (figure 10.86). Lime mortar, which in earlier centuries was expensive and not very adhesive, was only used for the thin final finishing coat.

Relatively little is known about the precise method of application. Most probably simple battens or trowels were used. According to eyewitness accounts, simple variants of earth plaster trowels existed in the period after 1945 of a kind much like those available today in Japan (section 4.5.4).

Further treatments

Unlike Japan for example, a tradition of the aesthetic surface treatment of earth plaster surfaces did not develop in Central Europe. After simple finishing of the surface, the plaster surface was general whitewashed. The purpose of the limewash was presumably to improve the hardness of the surface as well as to make room lighter when there were only small windows. Similarly the fineness of the lime material and visual clean lines of the white limewash application would have been reasons for its use. In cowsheds, the limewash also served to protect the health of the animals [Geinitz, 1822].

For the simple limewashing, the surface was coated several times with a whitewash or a mixture of lime and whey. Distempers of lime and fine sand were either painted on or rubbed into the still wet earth surface. Thin coats of lime mortar were also sometimes applied as plaster. The distemper or plaster was likewise given a limewash and it is not rare to find multiple layers of lime on top of one another.

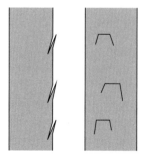

Fig. 10.85 : Diagram of timber surface worked with an axe

Fig. 10.86 :
Stucco on earth plaster in Brauweiler Abbey (Cologne Lowlands)

The working in of lime slurries as preparation for receiving a top coat of paint is described in the literature after 1945 for external render surfaces (section 10.3.1.4) [Niemeyer, 1946]. DIN 1169 *Lehmmörtel für Mauerwerk und Putz (Earth mortars for masonry and plaster)*, issued in June 1947, also describes surface stabilisation with a slurry of rich earth but concedes a preference for white lime and fine sand [DIN, 1947].

### 10.6.2 The repair of historical earth plasters

Earth plasters cannot be saved when they adhere so badly to the background that they can be removed by hand. Large sections of plaster that has parted from the background should be removed. They are easily diagnosed because they give clearly when pressed with a fingertip. Note that knocking is not a reliable test as knocking on solid topcoat plasters that are plastered over softer undercoats can also sound hollow although the adhesion is adequate. In general, when assessing earth plasters it is important to remember that by today's standards historical earth plasters have a low strength. Historical building materials should therefore be assessed according to a benchmark of their own.

The adhesion of layers of old earth plasters to one another is generally very good. Its adhesion to timber surfaces by contrast is not, as discussed above (figure 10.85). For this reason a plaster lath should be used (section 4.3.4).

When filling holes or repairing plaster surfaces, one should be aware that old earth plasters were often very lean. Mortars for repairing existing plasters must as far as possible correspond to the existing plaster. Contact surfaces should first be brushed down and then carefully pre-wetted. Thereafter the intensive rubbing in of a coarse-grain earth plaster slurry or a not too stiff earth mortar is recommended. Before the application of a new layer, the old earth plaster surfaces should be prepared similarly (section 4.6.3).

Earth plaster surfaces may be worth conserving in their old form if, for example, they exhibit valuable colour treatments or for other reasons. There is no generally applicable technique for fixing surfaces in place that have parted from the background, for example with injection plasters. Thin slurries of fine-grain clay or earth mass, sometimes with added cellulose have proven useful in this respect. They should not harden much harder than the existing historical earth plaster. Instead of water, thinned alcohol can be used as it evaporates more quickly.

### 10.6.3 Building material and building element properties

The building material and building element properties correspond to the details given for earth plasters in section 4.9.

# 11 Building legislation and business practice

## 11.1 Earth building regulations

In October 1994, the "Lehmbauordnung" (*Earth Building Ordinance*) came into force in the German Reich. It arose out of a need to regulate earth construction initially in the occupied zones in the East. In 1947 it was supplemented by the *Implementing Regulations for the Earth Building Ordinance* which covered all the occupied zones. Also in 1947 the DIN 1169 *Earth Mortars for Masonry and Plaster* was issued.

After the founding of the two separate German states in 1949, each state developed their own regulatory framework.

In the *Federal Republic of Germany*, the DIN 18951 *Building Codes for Earth Construction* were introduced in 1951. *Part 1*, the *Regulations for Construction Works* was a word-for-word reprint of the Earth Building Ordinance from 1944, its accompanying *Commentary (Part 2)* of the Implementing Regulations from 1947. The draft norms for the DIN 18952 – 18957, published in 1956, never progressed beyond preliminary norm status: they were not elaborated any further and never came into force. In 1971, the DIN 18951 was withdrawn.

In the *German Democratic Republic*, the regulations introduced for the occupied zones and the Federal Republic remained in force until 1951. In 1953 the DIN 18951 was replaced with the *Terms, Application and Processing of Earth Building Materials (Earth Building Code)*. Further codes followed including the *Regulations for the Application of Earth Construction* and the *Instructions for the Regulations for the Application of Earth Construction*, both from 1953. Exactly when these regulations ceased to be applicable is not known [Rath, 2004].

In the 1980s and the early 1990s, new developments in earth construction in Germany lacked a clear regulatory framework. In 1982, in response to a formal enquiry by Franz Volhard, the Ministry of the Interior of the Federal State of Hesse as the highest level planning authority, replied that "earth construction is recognised as long as its application corresponds to the cases described in the withdrawn norm," and that, "a special permit is not required providing that other applicable building control regulations and the regulations given in the norms are fulfilled." From the viewpoint of building control legislation, it was therefore possible to refer back to the withdrawn norm. At the time, however, the burgeoning modern earth construction movement had little in common with the building techniques of the 1940s. For historic monuments the local state building regulations also provide a formal procedure for *special consent on an individual basis*.

In 1996 the Building Authority in the north German state of Mecklenburg-Vorpommern identified an increasing need to regulate earth construction. The *German Institute of Building Technology (DIBt)* invited the *Dachverband Lehm e. V.* as the relevant professional association to take part in the drawing up of new building codes. In the view of those involved it was not practicable to simply revise the old DIN norms or to formulate new norms. Instead a new set of general regulations was drawn up reflecting the state of the art of earth construction. The actual process of formulating the norms was conducted by the *Dachverband Lehm e. V.* in 1997 and special attention was given to ensuring it had broad and democratic legitimation.

On the 23/24 September 1998, in a meeting of the *Conference of State Building Ministers (ARGE Bau)*, the new *Lehmbau Regeln* joined the official list of technical building regulations administered by the DIBt. The application of the *Lehmbau Regeln* is limited to buildings with up to two storeys and no more than two units (dwellings). In all other situations, special approval is required from the building control authorities. With regard to technical aspects, applications for special approval can refer to the *Lehmbau Regeln*. For comparatively minor applications, such as the interior plastering of residential buildings with more than two units, obtaining approval is mostly a formality. For structurally relevant applications, for example multistory loadbearing earth constructions, it is essential to clarify the question of building control approval at an early stage.

Inclusion in the official list of technical regulations (*Musterliste der Technischen Baubestimmungen*) conferred the *Lehmbau Regeln* recommended status for inclusion in the respective local building regulations. In April 2008, 14 of the 16 German states followed the recommendation; in the remaining states special consent is possible on an individual basis with reference to the Lehmbau Regeln.

In 2005, the *Dachverband Lehm e. V.* saw the need for a revision of the *Lehmbau Regeln*. In a second consultative process, similar in structure to the first, the results of almost ten years of practical application and new developments were incorporated into the regulations. The revised regulations were again included in the official list of technical regulations administered by the *DIBt* and the revised edition of the *Lehmbau Regeln* (3rd edition) was published in 2009. The new stipulation that the key constituents of industrially manufactured products be declared prepares the ground for the long-term development of norms compatible with European regulations, as recommended by the building control authorities.

In the meantime, the *Dachverband Lehm e.V.* has begun issuing *Technical Information Sheets*, for example the *TM01 – Requirements of earth plasters* in September 2008. In 2009, testing began at the *Federal Institute for Materials Research and Testing (BAM)* for the development of draft norms for testing soils for construction as well as earth mortars and earth blocks. In addition the use and application of earth mortars with reference to the DIN 18550, *Plastering/rendering and plastering/rendering systems – Execution*, and of earth blocks with reference to the DIN 1053-1, *Masonry – Part 1: Design and Construction*, is also planned.

## 11.2 Building trades and earth building qualifications

Earth construction works are generally regarded as a subset of the work undertaken by bricklayers and concrete workers. This is due to the traditional perception of earth as a massive construction method. Adobe, cob, mudwall and rammed earth are building shell constructions that are usually carried out by bricklayers or by unskilled labourers where specific skills are not required. Other construction work such as straw-clay construction or plastering were probably not specific enough to be attributed to a particular building trade.

Today, earth construction works are undertaken by the building trade normally associated with the respective task, even when usually undertaken with other materials. Consequently, plasterers apply earth plasters, drywall contractors apply clay panels and so on.

In 2001, the *Dachverband Lehm e.V.* implemented a further initiative with a view to establishing vocational qualifications for working with earth building materials. The *Specialist for Building with Earth (Fachkraft Lehmbau/FKL)* is a vocational training course aimed at those who already have a qualification or have completed their apprenticeship in a relevant building trade. Those from other branches may also be admitted providing they can demonstrate sufficient experience in building with earth.

The final examination is supervised by the local Chamber of Crafts and Trade (Handwerkskammer) and successful participants can then apply to be included in the register of qualified craftsmen. Companies offering earth construction services are classified in an own sub-category of Section A (Bricklayers and Concrete Workers). This entitles the holder to establish a building firm offering earth construction services (only). A list of qualified *DVL Earth Builders* is published online by the *Dachverband Lehm e.V.* The association is also involved in establishing the incorporation of earth building skills in the education and training of craftsmen in apprenticeship.

When *inviting tenders* for earth construction works, it is important to note the stipulation in the *Lehmbau Regeln*, that "the construction of loadbearing earth walls must be conducted under the direction and supervision of a sufficiently experienced specialist for earth building". Note that successful completion of the *FKL Specialist for Earth Building* course is on its own not sufficient to qualify as "sufficiently experienced".

## 11.3 Calculating the cost of earth building works

As with other building works, the cost of earth building works are planned and calculated with the help of written specifications (see below). Manufacturers and other companies publish specifications for their products and there are numerous online sources of specifications in various data formats.

The *Lehmbau Regeln* details the allocation of individual earth building elements to the respective trades and DIN norms as given in the *General Technical Conditions and Rules for Building Works* (ATV) that constitute Part C of the VOB, the *German Construc-*

Table 11.1 : Earth building works and corresponding trades

| Sect. | Building element (Lehmbau Regeln) | Building trade (VOB) | Norm VOB/C |
|---|---|---|---|
| 4.1.3 | Earth block walls | Bricklayer | DIN 18330 |
| 4.1.4 | Rammed earth walls | (Reinforced) concrete worker | DIN 18331 |
| 4.1.5 | Weller/mudwall walls | Bricklayer | DIN 18330 |
| 4.2 | Vaulted constructions | Bricklayer | DIN 18330 |
| 4.3.1 | Infill for timber-frame constructions | Bricklayer | DIN 18330 |
| 4.3.2 | Non-loadbearing rammed earth walls | (Reinforced) concrete worker | DIN 18331 |
| 4.3.3 | Non-loadbearing masonry | Bricklayer | DIN 18330 |
| 4.3.4 | Light earth wall construction (wet) | (Reinforced) concrete worker | DIN 18331 |
| 4.3.5 | Clay panel wall constructions | Bricklayer | DIN 18330 |
| 4.3.6 | Dry stacked earth wall constructions | Bricklayer | DIN 18330 |
| 4.3.7 | Sprayed wall constructions | (Reinforced) concrete worker | DIN 18331 |
| 4.3.8 | Mortar-bonded insulation boards | Plaster and stucco masons | DIN 18350 |
| 4.4 | Timber joist ceiling | Bricklayer | DIN 18330 |
| 4.5 | Plaster | Plaster and stucco masons | DIN 18350 |
| 4.6 | Rammed earth floor | Screed layer | DIN 18353 |
| 4.7 | Dry lining | Drywall contractor | DIN 18340 |

tion *Contract Procedures*. These specify, for example, the boundary between *associated services* and *additional services* as well as standard procedures for measuring up and collating a *bill of quantities*.

Determining typical work times and cost comparisons for particular constructions is problematic for several reasons:

▶ Specific features of earth building materials, such as moisture and odour absorption, may not be offered by other building materials that are otherwise technically comparable. Likewise, it is not always possible to find an equivalent surface finish in another material.
▶ The amount of *available data* on the cost of earth building constructions is still comparatively limited. As a result mean values may be disproportionately influenced by extreme values.
▶ The *demands of earth construction works* can vary quite considerably from job to job. A great deal of earth building construction is concerned with either high-quality surface finishes or complex and small-scale conservation work. Straightforward constructions with large surfaces are cheaper to produce but in terms of the statistical mean, they are underrepresented.
▶ The range of firms offering earth building services differs more widely in structure than in other sectors. Alongside conventional building contractors, plasterers and painters, there are also many small independent firms producing creative and artistic work.

As a result, it is hard to determine practically-applicable typical values for work times and cost calculations. The following list of typical values should therefore be used with a degree of caution. The values given below are for general orientation only. It is important to conduct own timings and cost calculations as well as own research, as the prices of actual services can vary sometimes considerably.

## 11.3.1 Typical work times

The following list of earth constructions with typical work times covers a selection of the most important earth construction works. The tasks they constitute are defined in written specifications. When not otherwise stated, all values given are in minutes per m².

**Earth plasters: Plaster lath application, timber members**  50 min
(6 min/m at B = 12 cm)

The application of reed plaster lath to timber members in the plane of the applied plaster on walls or ceilings. Fixing with galvanized staples (L ≥ 25 mm) with the reed stems parallel to the timber beam.

### Earth plasters: Plaster lath application, large areas    11 - 16 min

Covering of wood or other wall and ceiling surfaces with reed plaster lath. Fixing with galvanized staples (L ≥ 25 mm). Reed stems should be arranged at right angles to sheathing, planking or similar.

### Earth plasters: Earth mortar spatter coat    6 - 8 min

Application of a spatter coat of course earth mortar onto walls/ceilings in preparation for a single-layer application of coarse earth plaster. Provides a key for the subsequent plaster layer and ensures consistent absorbency of the substrate. Includes preparation of the substrate (brushing down, wetting). Near 100 % coverage, textured surface finish.

### Earth plasters: Levelling course    6 - 8 min

Application of a levelling course of earth mortar onto walls/ceilings including preparation of the substrate (brushing down, wetting). Thickness: 15 mm.

### Earth plasters: Undercoat plaster, coarse earth mortar    13 - 17 min

Application of a single layer of plaster made of coarse earth mortar to walls/ceilings as undercoat including preparation of the substrate (brushing down, wetting). Surface quality level Q2, ready to receive subsequent 3 - 5 mm layer of earth plaster. Plaster thickness: 15 mm.

### Earth plasters: Supplement for preparation of coarse earth undercoat plaster for receiving self-coloured earth plaster    5 - 7 min

Supplement for fine rubbing down of an undercoat earth plaster on walls/ceilings to create a level and crack-free surface. Where necessary, application of a thin levelling layer of fine earth mortar. Surface finish according to quality level Q3, ready to receive a subsequent 2 mm layer of coloured earth plaster.

### Earth plasters: Plaster reinforcement, large area (flax mesh)    6 min

Embedding of a plaster reinforcement mesh (large area) in the fresh, wet surface of an earth undercoat plaster on walls/ceilings including necessary overlapping.

### Earth plasters: Topcoat plaster / single layer plaster of coarse earth mortar    14 - 19 min

Application of a layer of plaster of coarse grain earth plaster to walls/ceilings as topcoat plaster or single layer application. Including necessary preparation of the substrate (brushing down, wetting). Surface finish according to quality level Q3, ready for painting. Plaster thickness: 10 mm.

### Earth plasters: Topcoat plaster / single layer plaster of fine earth mortar — 12 - 17 min

Application of a plaster layer of fine earth mortar onto walls/ceilings as topcoat plaster or single layer application. Including necessary preparation of the substrate (brushing down, wetting). Surface finish according to quality level Q3, ready for painting. Plaster thickness: 3 - 5 mm.

### Earth plasters: Corner bead — 7 min/m

Appropriate and sufficiently anchored fixing of a galvanized/stainless steel/ plastic corner bead with bonding compound.

### Earth plasters: Rounding off edges — 15 min/m

Careful rounding of corners and edges of walls/ceilings without a corner bead. Including application of strips of reinforcement mesh as required. Corner radius of up to 10 mm.

### Earth plasters: Supplement for special surface finish (smoothed) — 5 - 7 min

Supplement for earth plaster surfaces on walls/ceilings with a special surface finish. Surface quality according to Q4, smoothed, ready to receive paint finish.

### Earth plasters: Brush-on primer application — 3 - 5 min

Application of a primer with fine grain component to walls or ceilings. Working according to manufacturer's recommendations. Even grain texture with a good key for subsequent plastering with a coloured earth plaster mortar or brush-on earth plaster.

### Earth plasters: Topcoat plaster / single layer of coloured earth plaster — 19 - 25 min

Application of a layer of coloured earth plaster to walls or ceilings as a topcoat plaster or single layer application with a thickness of 2 mm on suitable substrates (earth plaster surfaces should accord to quality level Q3, rubbed). Including any necessary preparation of the substrate (brushing down, wetting). Rubbed surface finish as facing finish.

### Earth plasters: Supplement for brushing / washing-down surface — 5 - 7 min

Supplement for brushing or washing down coloured earth plaster surfaces after full drying. Removal of residues through washing down with a soft, moist, clean sponge (2 - 3 strokes)/burnishing lightly with a soft wallpaper brush or similar.

### Earth finishes: Application of a coat of earth paint or brush-on earth plaster — 4 - 7 min

Application of a brush-on earth plaster or earth paint onto walls or ceilings. Working of material according to manufacturer's recommendations with a painter's brush or roller.

### Dry earth construction: Dry stacked wall lining — 25-30 min

Installation of a batten at the base of the wall. Insertion of earth bricks in stacked bond. Firm clamping of the bricks with a batten every third/fourth course. Fixing of the battens to the existing construction with appropriate screws. Insertion of a clamping batten at the top. Including cutting blocks to size at the sides and top or alternatively filling smaller gaps with earth masonry mortar. Finished result ready for cladding with clay plasterboard or for plastering over. Thickness: DF 52 mm.

### Dry earth construction: Subconstruction for clay plasterboard (wall) — 20 min

Manufacture of a wall stud construction out of softwood studs/metal profiles ready for cladding with clay plasterboard. Sufficiently stable construction including fixing to the existing structure.

### Dry earth construction: Cladding of wall surfaces with clay plasterboard — 20 min

Cladding of wall surfaces with clay plasterboard. Mounting of the boards in staggered bond including fixing with appropriate screws. Finished result ready for application of scrim tape and joint filling. Board thickness: 20 mm/25 mm.

### Dry earth construction: Reinforcement of clay plasterboard panel joints — 6 min

Reinforcement of joints between clay plasterboard panels used as cladding for walls or ceilings with mesh strips. Embedding in earth mortar or brushed into place with an earth slurry. Ready to receive a subsequent 3-5 mm thick plaster layer.

### Dry earth construction: Clay plaster panels as dry plaster on a mineral substrate — 16 min

Cladding of walls, ceilings or roof inclines with clay plaster panels as dry plaster. Full-surface bonding to the substrate in a bed of earth adhesive. For very uneven surfaces and ceilings and roof inclines, additional anchoring of the panels with lightweight board pins or impact dowels. Finished result ready for application of reinforcement and joint compound. Panel thickness: 16 mm.

### Internal insulation: Construction of reed plaster lath formwork for light earth, including filling with light earth — 50-70 min

Erection of a construction of vertical battens on the inner face of an external wall at 30-35 cm centres. Parallel and vertical alignment of the battens as per plan. Stable anchorage of the battens with the existing construction including prevention against buckling. Successive covering of the battens with reed plaster lath as the cavity is filled with light earth mixture. Fixing with galvanized wire (at least 1.2 mm) and staples (L ≥ 25 mm). Even filling of the light earth mixture into the formwork cavity, taking care to ensure all holes are filled. Finished result ready for plastering. Wall thickness: 15 cm.

**Internal insulation: Creation of openings in the light earth formwork for windows and doors, including covering** 45 - 60 min

Measuring out and construction of the window/door jambs according to plan out of battens. Measuring out and construction of the lintel out of scantling(s). Insertion of formwork for jambs and lintel depths > 6 cm (incl. subsequent removal) or alternatively mounting of permanent formwork made of an appropriate plaster lath. Finished result ready for plastering. Width of opening: ca. 1 m.

**Internal insulation: Earth block masonry wall lining** 70 - 85 min

Construction of a half-brick thick masonry wall lining (11.5 cm) out of light earth blocks and earth masonry mortar. The wall lining should lie ca. 1 cm forward of the existing external wall. Backfilling of the resulting cavity with light earth masonry mortar during the construction process. Including delivery and mounting of perforated steel strip and screws/galvanized wall ties for firmly anchoring the wall lining with the external wall during filling of the cavity as well as for structural stability. Finished result ready for plastering.

**Internal insulation: Wall lining with insulation board and earth mortar** 20 - 28 min

Careful preparation of the substrate (brushing down, wetting), thereafter application of max. 1 cm thick layer of earth undercoat plaster to the full surface of the inner face of an external wall. Application of the insulation board by pressing firmly into the plastic earth mortar. Alternatively, attach the insulation board with earth adhesive applied with a notched spatula to its full surface. Additional pressure through the use of sufficiently long screws, impact dowels or lightweight board pins, approx. 5 fixing points per m². The bond between insulation board and mortar must be firm and durable across its entire surface. Finished result ready for plastering. Softwood insulation board thickness: 60 mm.

**Masonry: Earth block masonry with 2DF blocks** 48 - 92 min

Construction of a half-brick thick (11.5 cm) wall of earth blocks with earth masonry mortar. Blocks should laid in brick bond taking care to fill joins fully. Wall construction up to a height of 2.0 m, in exceptions a full storey, per day. Wall must be structurally stable, including necessary connections to the existing structure. Including any necessary protection against the weather during construction. Finished result ready for plastering.

### Masonry: Earth block masonry wall infill of a timber-frame construction     58 - 102 min

Construction of half-brick thick (11.5 cm) sections of wall infill made of earth blocks with earth masonry mortar. Blocks should be laid in brick bond, taking care to fill joins fully. Anchoring of the section of masonry with the timber frame through the insertion of triangular battens. Including possible necessary stabilisation for larger panels in the form of horizontal battens or boards. Finished result ready for plastering.

### Masonry construction: Supplement for facing blockwork     15 min

Supplement for the careful execution of aforementioned masonry in a quality suitable for facing blockwork.

### Masonry construction: Creation and covering over of openings     20 - 25 min

Measuring out and creation of masonry junctions for window and door jambs/rebates for receiving door frames. Measuring out and creation of lintels from scantlings. Width of opening: ca. 1 m.

### Rammed earth: Rammed earth wall with exposed surface     350 - 1250 min

Construction of a flat and sufficiently stable and rigid formwork (to withstand a surface load of 60 kN/m²). Checking the consistency and homogeneity of the rammed earth mixture. Introduction of the rammed mixture into the formwork in 10 - 15 cm thick layers. Repeated careful and uniform ramming of the rammed earth with a hand-operated or mechanically-powered tamper. Structurally stable realisation including any necessary anchorage to the existing building shell. Careful execution of the building works according to design plan so that the resulting face is of sufficient quality to remain exposed to view for the long-term. Removal of the shuttering formwork. If required careful highlighting of the coarser aggregate by removing smaller particles with a wire brush or similar, either by hand or machine. Timing and kind of working of the surface according to the design plan. Including any necessary protection against the weather and frost during construction as well as outside of construction times until handover of the finished work. Wall thickness: 30 cm.

### Rammed earth: Rammed earth floor     200 - 350 min

Checking of the stability and suitability of the underlying surface. Checking of the consistency and homogeneity of the rammed earth mixture. Introduction of the rammed earth mixture in several layers of appropriate thickness. Repeated careful and uniform ramming by hand or machine. Incorporation of a wide mesh geogitter in the upper layer of the rammed floor construction. Where necessary, insertion of an edging strip along the walls to minimise the risk of crack formation on the floor's surface. Surface treatment according to design stipulations for an exposed rammed earth floor, i.e. without any further floor covering.

### Timber-frame wall infill: Repair of an existing external wall panel infill with straw-clay application    65 - 80 min

Inspection of all existing infill panels. Where necessary, repair of individual staves or wattle. Removal of previous repairs made with unsuitable materials and loose sections of earth, including correct disposal. Roughening of surfaces. Opening of the joint around the infill panel. Removal of dust, dirt and adhesion-impairing coatings. After careful pre-wetting of the earth surfaces, application of a 2 - 3 cm thick layer of straw-clay mass, leaned down if necessary, including filling of holes and the joint around the infill panel. Second application of a straw-clay layer to a depth of 15 - 18 mm short of the face of the timber members to leave room for a subsequent facing plaster of two layers of lime plaster. Preparation of the plaster substrate for the lime plaster by careful roughening (stippling) of the fresh straw clay with a plaster scraper or similar.

### Timber-frame wall infill: Preparation of an existing external wall panel infill to receive a single layer of lime external render    15 - 22 min

Inspection of all existing infill panels. Where necessary removal of previous repairs made with unsuitable materials and loose sections of earth, including correct disposal. Removal of dust, dirt and adhesion-impairing coatings. Roughening of the surface, for example with a plaster scraper. After careful pre-wetting, preparation of the earth surfaces with coarse earth mortar including filling of holes. Finished result ready to receive a 4 - 6 mm thick single layer of lime external render.

### Timber-frame wall infill: Wattle and daub infill of an external wall panel    120 - 135 min

Cutting of staves to fit in existing grooves in the beams. Weaving of willow rods around the staves to form a wattle. Application of a layer of straw-clay to the outside and inside of the wattle. On the external face, straw-clay application up to a depth of 15 - 18 mm short of the face of the timber members to leave room for a subsequent facing plaster of two layers of lime plaster. Preparation of the plaster substrate for the lime plaster by careful roughening (stippling) of the fresh straw clay with a plaster scraper or similar. On the internal face, simple smooth surface finish sufficient to fully cover the staves and wattle.

### Timber-frame wall infill: Light earth block infill of external wall panel    70 - 115 min

Construction of half-brick thick (11.5 cm) sections of wall infill made of earth blocks with earth masonry mortar. Blocks should be laid in brick bond, taking care to fill joins fully. Anchoring of the section of masonry with the timber frame through the insertion of triangular battens. External face of masonry should lie 15 - 18 mm behind the face of the timber members to leave room for a subsequent facing plaster of two layers of lime plaster. Scratching out of horizontal and vertical mortar joints (full width, not v-shaped) in fresh masonry walling to a depth of 0.5-1.0 cm to provide a mechanical key for the subsequent lime plaster.

**Timber-frame wall infill: Air-hardening coarse lime plaster with added hair as undercoat render** — 20-25 min

Spray mist pre-wetting of substrate directly in advance of plastering (panel for panel), where necessary repeatedly. Application of lime render as undercoat with plasterer's darby, alternatively vigorous application with a trowel. Creation of a roughened surface suitable for receiving subsequent topcoat render. Render thickness: 8/10 mm.

**Timber-frame wall infill: Air-hardening fine-finish lime as topcoat render** — 20-25 min

Spray mist pre-wetting of substrate directly in advance of plastering (panel for panel), where necessary repeatedly. Application of topcoat of lime render. Render surface should be flat and finish in line with the front face of the timber members. Fine rubbed/smoothed surface finish ready for painting with lime wash. Render thickness: 4-6 mm.

## 11.3.2 Calculating typical constructions

Table 11.2 : A comparison of the cost of typical constructions

| Earth construction | Alternative 1 | Alternative 2 (simple) |
|---|---|---|
| **PLASTERS, WALL COVERINGS, PAINTS** | | |
| Earth plaster, 2 coats<br>+ primer coat<br>+ 2 coats brush-on earth plaster<br>32.50–37.00 €/m² | Natural lime plaster, 2 coats<br>+ 2 coats of lime wash<br><br>30.50–32.50 €/m² | Gypsum plaster, 1 coat<br>on a pre-sprayed or primer coat<br>+ wood-chip wallpaper<br>+ 2 coats of dispersion paint<br>20.50–24.50 €/m² |
| Earth plaster, 1 coat<br>+ primer coat<br>+ 2 coats brush-on earth plaster<br>20.00–25.50 €/m² | Lime plaster, 1 coat<br>+ 2 coats of lime wash<br><br>17.50–23.50 €/m² | Gypsum plaster, 1 coat<br>+ wood-chip wallpaper<br>+ 2 coats of dispersion paint<br>16.50–21.50 €/m² |
| Coloured earth plaster, thin coat<br>on a primer coat<br>19.00–24.00 €/m² | Synthetic roll-on plaster, thin coat<br>on a primer coat<br>16.50–19.50 €/m² | Glass-fibre wallpaper<br>+ 2 coats of dispersion paint<br>11.00–15.50 €/m² |
| 2 coats of brush-on earth plaster<br>on a primer coat<br>8.00–12.00 €/m² | 2 coats of brush-on lime plaster<br>on a primer coat<br>8.00–12.00 €/m² | Cellulose brush-on plaster<br>or cotton brush-on plaster<br>10.50–14.50 €/m² |
| Rounding of corners and edges<br>11.50–16.50 €/m | | Edging with corner bead<br>6.50–10.00 €/m |

Table 11.2 : continued

| Earth construction | Alternative 1 | Alternative 2 (simple) |
|---|---|---|
| **DRYWALL CONSTRUCTION** | | |
| **Wall with dry stacked panel infill**<br>Timber stud wall construction with earth block dry sacked infill<br>+ each side: clay board 20 mm<br>+ each side: topcoat plaster, reinf.<br>+ each side: primer coat<br>+ each side: 2 × brush-on earth plaster<br>173.00–185.00 €/m² | **Dry stacked panel infill**<br>Timber stud wall construction with stacked panel infill (any material)<br>+ each side: 2 × gypsum pb 9.5 mm, joint filler, reinforced<br>+ each side: wood-chip wallpaper<br>+ each side: 2 × dispersion paint<br>128.00–142.00 €/m² | **Panel infill with insulation**<br>Metal stud construction with mineral wool<br>+ each side: gypsum pb 12.5 mm, joint filler, reinforced<br>+ each side: wood-chip wallpaper<br>+ each side: 2 × dispersion paint<br>68.00–80.00 €/m² |
| **Dry stacked wall lining**<br>Battens/subconstruction and earth block dry stacked wall lining, clay board 20 mm<br>+ topcoat plaster, reinforced<br>+ primer coat<br>+ 2 coats of brush-on earth plaster<br>74.80–83.30 €/m² | **Wall lining**<br>Battens/subconstruction and gypsum plasterboard 12.5 mm, joint filler, reinforced<br>+ wood-chip wallpaper<br>+ 2 coats of dispersion paint<br>35.60–44.30 €/m² | **Wall lining**<br>Battens/subconstruction and 2 × gypsum plasterboard 9.5 mm, joint filler, reinforced<br>+ wood-chip wallpaper<br>+ 2 coats of dispersion paint<br>39.60–46.70 €/m² |
| **Clay plasterboard partition wall**<br>Battens/subconstruction and cellulose insulation<br>+ each side: clay board 20 mm<br>+ each side: topcoat plaster, reinf.<br>+ each side: primer coat<br>+ 2 coats of brush-on earth plaster<br>129.00–146.00 €/m² | | **Partition wall**<br>(as 6 but with additional insulation) |
| **Clay plaster panels**<br>clay plaster panels, 16 mm (glued)<br>+ topcoat plaster, reinforced<br>+ primer coat<br>+ 2 coats of brush-on earth plaster<br>49.50–56.00 €/m² | | **Plasterboard dry lining**<br>gypsum pb 12.5 mm + bond. compound joint filler, reinforced<br>+ wood-chip wallpaper<br>+ 2 coats of dispersion paint<br>27.00–34.00 €/m² |
| **MASONRY PANEL INFILL OF EXPOSED TIMBER-FRAME CONSTRUCTION** | | |
| **Earth block masonry 11.5 cm**<br>+ each side: 2 coats of earth plaster<br>+ each side: primer coat<br>+ each side: 2 × brush-on earth plaster<br>112.50–137.50 €/m² | **Brick masonry 11.5 cm**<br>+ 2 coats of lime plaster<br>+ 2 coats of lime paint<br>92.50–109.00 €/m² | **Aerated concrete block 11.5 cm**<br>+ each side: joint compound layer<br>+ each side: 2 × dispersion paint<br>57.50–63.00 €/m² |
| **INTERNAL INSULATION LAYER** | | |
| **Light earth wall lining 15 cm**<br>Battens, rolled reed plaster lath and wood-chip light earth 15 cm<br>+ 2 coats of earth plaster, reinforced<br>+ primer coat<br>+ 2 coats of brush-on earth plaster<br>108.00–115.00 €/m² | | **Stud wall, mineral wool 10 cm**<br>Timber stud wall construction with mineral wool<br>+ vapour barrier<br>+ gypsum plasterboard 12.5 mm, joint filler, reinforced<br>+ wood-chip wallpaper<br>+ 2 coats of dispersion paint<br>59.50–70.00 €/m² |

Table 11.2 : continued

| Earth construction | Alternative 1 | Alternative 2 (simple) |
|---|---|---|
| **Earth block masonry 11.5 cm** with earth mortar cavity fill + 2 coats of earth plaster + primer coat + 2 coats of brush-on earth plaster 95.00–115.00 €/m² | **Perforated brick masonry 11.5 cm** with cavity filling (any material) + 2 coats of lime plaster + 2 coats of lime paint 75.00–85.00 €/m² | **Aerated concrete block 11.5 cm** no cavity filling (air cavity) + joint compound layer + wood-chip wallpaper + 2 coats of dispersion paint 58.50–75.00 €/m² |
| **Softwood insulation board 6 cm** on earth plaster levelling course + topcoat plaster, reinforced + primer coat + 2 coats of brush-on earth plaster 80.00–87.00 €/m² | **Calcium silicate board 5 cm** bonded + topcoat plaster, reinforced + primer coat + 2 coats of brush-on earth plaster 85.00–95.00 €/m² | **Battens, mineral wool 6 cm** + vapour barrier + gypsum wallboard 12.5 mm + joint filler, reinforced + wood-chip wallpaper + 2 coats of dispersion paint 54.50–61.50 €/m² |
| **BRICKWORK INFILL OF HISTORICAL EXPOSED TIMBER-FRAME FAÇADE** | | |
| **Earth block masonry 11.5 cm** + 2 coats of lime plaster + 2 coats of lime paint 98.00–120.00 €/m² | | **Perforated brick masonry 11.5 cm** + 2 coats of lime plaster + 2 coats of lime paint 98.00–120.00 €/m² |
| **RAMMED EARTH** | | |
| **Exposed rammed earth wall 30 cm** 300.00–900.00 €/m² | **Exposed stone wall 30 cm** 300.00 €/m² **Exposed earth block wall 36.5 cm** 135.00–155.00 €/m² | **Exposed concrete wall 30 cm** simple quality 130.00–150.00 €/m² |
| **Rammed earth floor 10-15 cm** waxed surface + geogitter reinforcement 150.00 €/m² | | **Cement screed 6-8 cm** waxed surface 45.00 €/m² |

*Boundary conditions: Prices for construction works in Germany as of 2010 excl. VAT; realistic upwards of a quantity of 100 m²; based on a labour cost of 40 € per hour. Prices given are for wall surfaces; for ceiling surfaces add 15-20 %. With another paint/coating material, a further reduction of the cost of earth constructions is possible (by approx. 2 €).*

# Bibliography

## Chapter 2 – The raw material – soils for construction

| | |
|---|---|
| DIN EN 1015-3:1999 | Prüfverfahren für Mörtel für Mauerwerk. Teil 3: Bestimmung der Konsistenz von Frischmörtel (mit Ausbreittisch) |
| DIN EN 1015-11:2007 | Prüfverfahren für Mörtel für Mauerwerk. Teil 11: Bestimmung der Biegezug- und Druckfestigkeit von Festmörtel |
| DIN 4022-1:1987 | Benennen und Beschreiben von Boden und Fels. |
| DIN 18123:1996-11 | Baugrund, Untersuchung von Bodenproben – Bestimmung der Korngrößenverteilung |
| [Dierks/Ziegert, 2002] | Dierks, K., Ziegert, C.: Neue Untersuchungen zum Materialverhalten von Stampflehm. In: Steingass, P.: Moderner Lehmbau 2002. Tagungsband, Fraunhofer IRB-Verlag, Stuttgart 2002 |
| [Dierks/Ziegert, 2000] | Dierks, K., Ziegert, C.: Materialprüfung und Begleitforschung im tragenden Lehmbau. In: Wiese, K. (Ed.): Lehm 2000. Tagungsband, Fraunhofer IRB-Verlag, Stuttgart 2000 |
| [Heim, 1990] | Heim, D.: Tone und Tonminerale, Spektrum Akademischer Verlag, Stuttgart 1990 |
| [Müller, 2007] | Müller, U.: Untersuchung zum Geruchsabsorptionsverhalten von Lehmputzen. Research project commissioned by Claytec, Bundesanstalt für Materialforschung- und Prüfung (BAM), FG VII.1 Baustoffe, unveröffentlichter Abschlussbericht, Berlin 2007 |
| [Jasmund/Lagaly, 1993] | Jasmund, K., Lagaly, G.: Tonminerale und Tone. Steinkopff-Verlag, Darmstadt 1993 |

## Chapter 3 – Earth building materials: composition & properties

[Goris, 2008]  Goris, A. (Ed.): Bautabellen für Architekten. Werner-Verlag, Cologne 2008

[Dettmering / Kollmann, 2001]  Dettmering, T., Kollmann, H.: Putze in Bausanierung und Denkmalpflege. Verlag für Bauwesen, Berlin 2001

## Chapter 4 – Earth plasters

[Müller, 2007]  Müller, U.: Untersuchung zum Geruchsabsorptionsverhalten von Lehmputzen. Auftragsforschung der Firma Claytec, Bundesanstalt für Materialforschung- und Prüfung (BAM), FG VII.1 Baustoffe, unveröffentlichter Abschlussbericht, Berlin 2007

[DVL, 2009]  Dachverband Lehm e. V. (Ed.): Lehmbau Regeln 2009, Vieweg + Teubner, Wiesbaden 2009

[DIN, 2002]  Deutsches Institut für Normung e. V. (Ed.): Vergabe- und Vertragsordnung für Bauleistungen, Beuth Verlag, Berlin 2002

DIN V 18550 2005-04  Putz und Putzsysteme – Ausführung (preliminary norm)

[Franz / Schwarz / Weißert, 2008]  Franz, R., Schwarz, E., Weißert, M.: Kommentar ATV DIN 18350 und DIN 18299 Putz- und Stuckarbeiten, Vieweg + Teubner, Wiesbaden 2008

[GIPS, 2003]  Bundesverband der Gipsindustrie e. V. (Ed.): Putzoberflächen im Innenbereich, Merkblatt, 2003

[GIPS, 2007]  Bundesverband der Gipsindustrie e. V. (Ed.): Verspachtelung von Gipsplatten – Oberflächengüten, Merkblatt, 2007

[DVL 2008]  Dachverband Lehm e. V. (Ed.): Technische Merkblätter Lehmbau: – Blatt 01: Anforderungen an Lehmputze, DVL 2008

[Röhlen, 2008]  Röhlen, U.: Auftreten von Schimmelpilzen auf Lehmputzoberflächen während der Trocknung: Ein baustoffspezifisches Problem? In: Berufsverband Deutscher Baubiologen e. V. (Ed.): 12. Pilztagung des VDB. Tagungsband, Verlag des AnBUS e. V., Fürth 2008

[Natureplus, 2005]  Natureplus e. V. (Ed.): Vergaberichtlinie 0803 Lehmputzmörtel, Neckargemünd 2005

| | |
|---|---|
| DIN EN 998-1 2003-09 | Festlegungen für Mörtel im Mauerwerksbau, Teil 1: Putzmörtel |
| DIN 4108-3 2001-07 | Klimabedingter Feuchteschutz, Anforderungen, Berechnungsverfahren und Hinweise für Planung und Ausführung |
| DIN 4102-4 1994-03 | Brandverhalten von Baustoffen und Bauteilen, Zusammenstellung und Anwendung klassifizierter Baustoffe, Bauteile und Sonderbauteile. |

## Chapter 7 – Internal insulation with earth materials

| | |
|---|---|
| [WTA, 2003] | Wissenschaftlich-Technische Arbeitsgemeinschaft für Bauwerkserhaltung und Denkmalpflege e.V. (Ed.): Fachwerkinstandsetzung nach WTA I: Bauphysikalische Anforderungen an Fachwerkgebäude, WTA Publications, Munich 2003 |
| [Volhard, 1995] | Volhard, F.: Leichtlehmbau, alter Baustoff – neue Technik, Verlag C. F. Müller, Karlsruhe 1995 |

## Chapter 8 – Earth block masonry

| | |
|---|---|
| [DVL, 2011] | Dachverband Lehm e.V. (Ed.): Technisches Merkblatt Lehmsteine. DVL Weimar, scheduled for 2011 |
| DIN 1053-1:1996-11 | Mauerwerk – Teil 1: Berechnung und Ausführung |
| DIN 1055-100:2007-09 | Mauerwerk – Teil 100: Berechnung auf der Grundlage des semiprobabilistischen Sicherheitskonzepts |
| DIN 4102-1:1998-05 | Brandverhalten von Baustoffen und Bauteilen – Teil 1: Baustoffe; Begriffe, Anforderungen und Prüfungen |
| DIN 4103-1:1984-07 | Nichttragende innere Trennwände; Anforderungen, Nachweise |

## Chapter 9 – Rammed earth construction

| | |
|---|---|
| [Ziegert, 2006] | Ziegert, C.: Orientierende Untersuchungen zu Befestigungsmitteln in Stampflehmmischungen der Fa. Claytec. Study undertaken in conjunction with HILTI Germany for Claytec e.K., Berlin, 2006 |

## Chapter 10.2 – Solid earth wall constructions

[Bauwelt, 1949] Bauwelt: Stundenaufwand für Lehmbauarbeiten. Neue Bauwelt 1949, No. 36, Munich 1949

[Bosslet, 1920] Bosslet, A.: Alte und neue Lehmbauten. In.: Zeitschrift für Wohnungswesen in Bayern, No. 6/8 – 1920, Munich 1920

[Düttmann, 1920] Düttmann, W.: Der Lehmbau in künstlerischer Beziehung. In: Die Bauwelt 34/1920, Munich 1920

[Geinitz, 1822] Geinitz, L.: Über den Lehmbau im Altenburgischen. In: Monatsblatt für Bauwesen und Landesverschönerung, No. 3/1822, Munich 1822

[Fauth, 1933] Fauth, W.: Der Lehm als Baustoff. Vol. 1, Eberswalde – Berlin 1933

[Fiedler, 1965] Fiedler, A.: Kursächsische Landesverordnungen des 16.-18. Jh. und ihre Einwirkungen auf die ländliche Bauweise. In: Deutsches Jahrbuch für Volkskunde, Berlin 1965

[Grimm / Grimm, 1885] Grimm, J., Grimm, W.: Deutsches Wörterbuch. 6 volumes, Leipzig 1885

[Heimberg / Riesche, 1998] Heimberg U., Rieche, A.: Die römische Stadt – Planung, Architektur, Ausgrabung. Pulheim 1998

[Kienzle / Ziegert, 2008] Kienzle, P., Ziegert, C.: The reconstruction of Roman rammed earth buildings in the Archaeological Park in Xanten. In: LEHM 2008. DVL Weimar 2008

[Lange, 1779] Lange, J. G.: Zufällige Gedanken über die nothwendige und bequeme Wirthschaftliche Bauart auf dem Lande. J.F. Korn, Breslau 1779

[Miller / Grigutsch, 1947] Miller, T., Grigutsch, E.: Lehmbaufibel. Published by the Forschungsgemeinschaft Hochschule / Weimar, Weimar 1947

[Wrede, 1920] Wrede, E.: Die Lehmwellerwandbauweise. In: Der Siedler. Number 11/12, December 1920, Dresden 1920

[Wähler, 1940] Wähler, M.: Thüringische Volkskunde. Jena 1940

[Seebaß, 1803] Seebaß, C. L.: Die Pisé-Baukunst. German translation of the French original edition by Francois Cointereaux, Reprint of the original edition printed in 1803, Leipzig 1989

| | |
|---|---|
| [Simons, 1965] | Simons, G.: Bäuerliche Lehmgewinnung. Göttingen 1965 |
| [SMAD, 1949] | SMAD: Befehl 209 der Sowjetischen Militär-Administration vom 9.9.1947: Maßnahmen zum wirtschaftlichen Aufbau der Neubauernwirtschaften, Berlin 1949 |
| [ThürLandOrd, 1556] | Thüringische Landesverordung von 1556. Christian Rödinger in Jena, 1556. In: Schmolitzky, O.: Volkskunst in Thüringen vom 16. bis zum 19. Jahrhundert. Weimar 1964 |
| [Ziegert, 2003] | Ziegert, C.: Lehmwellerbau. Konstruktion, Schäden und Sanierung. Fraunhofer IRB-Verlag, Stuttgart 2003 |

## Chapter 10.3 – Timber-frame panel infill

| | |
|---|---|
| [Figgemeier, 1994] | Figgemeyer, M.: Der Baustoff Lehm im historischen Fachwerkbau, Dissertation at the Philipps-Universität Marburg, Marburg 1994 |
| [WTA, 1999] | Wissenschaftlich-Technische Arbeitsgemeinschaft für Bauwerkserhaltung und Denkmalpflege e.V. (Ed.): Fachwerkinstandsetzung nach WTA III: Ausfachung von Sichtfachwerk, WTA Publications, Munich 1999 |
| [Volhard, 1992] | Volhard, F.: Das Gotische Haus Römer 2-4-6, Strohlehmausfachungen und Lehmputze, Limburg a. d. Lahn, Forschungen zur Altstadt, Vol. 1, p. 212-224 / Magistrat der Stadt Limburg an der Lahn 1992 |
| [Niemeyer, 1946] | Niemeyer, R.: Der Lehmbau und seine praktische Anwendung, Ökobuch Verlag, Freiburg 1987 |
| [Breidenbach/ Röhlen, 2008] | Claytec e.K.: Architektenmappe – Die Anleitung zum modernen Lehmbau, Viersen 2008 |
| [Eckermann/Veit, 1996] | Eckermann, W., Veit, J.: Biologisch Bauen – Ein Nachschlagewerk für baubiologische und ökologische Planungen, Aktualisierung Mai 2004, Institut für Baubiologie GmbH Rosenheim (Ed.), Rosenheim 2004 |
| [Gaul, 2000] | Gaul, B.: Gefacheputz, Arbeitsblatt Deutsches Zentrum für Handwerk und Denkmalpflege, Probstei Johannesberg, Fulda 2000 |

| | |
|---|---|
| [WTA, 1999] | Wissenschaftlich-Technische Arbeitsgemeinschaft für Bauwerkserhaltung und Denkmalpflege e.V. (Ed.): Fachwerkinstandsetzung nach WTA IV: Beschichtungen auf Fachwerkwänden: Ausfachungen, Putze, WTA Publications, Munich 1999 |
| [DIN, 2002] | DIN 4108-4, 2002-02 / DIN Deutsches Institut für Normung / Beuth Verlag 2002 |
| [DIN, 1994] | DIN 4102-4, 1994-03 / DIN Deutsches Institut für Normung / Beuth Verlag 1994 |
| [DVL, 2009] | Dachverband Lehm e.V. (Ed.): Lehmbau Regeln 2009, Vieweg + Teubner, Wiesbaden 2009 |
| [DIN, 2005] | DIN V 18550, 2005-04 / DIN Deutsches Institut für Normung / Beuth Verlag 2005 |
| [WTA, 2004] | Wissenschaftlich-Technische Arbeitsgemeinschaft für Bauwerkserhaltung und Denkmalpflege e.V. (Ed.): Fachwerkinstandsetzung nach WTA XII: Brandschutz bei Fachwerkgebäuden, WTA Publications, Munich 2004 |

## Chapter 10.5 – Earth floors

| | |
|---|---|
| [Breymann, 1881] | Breymann, G. A.: Allgemeine Bau-Constructions-Lehre. In: I. Theil: Constructionen in Stein, Verlag J. M. Gebhardt, Leipzig 1881 |
| [DIN, 1965] | DIN 1965: Estrich- und Fliesenarbeiten. In: DIN Taschenbuch 3: Technische Vorschriften für Bauleistungen, Beuth Verlag, Berlin 1926 |

## Chapter 10.6 : Historical earth plasters

| | |
|---|---|
| [Niemeyer, 1946] | Niemeyer, R.: Der Lehmbau und seine praktische Anwendung, Ökobuch Verlag, Freiburg 1987 |
| [Pollack / Richter, 1952] | Pollack, E., Richter, E.: Technik des Lehmbaus – Grundlagen für Entwurfsbearbeitung, Bauleitung und Planung, Verlag Technik, Berlin 1952 |

[Geinitz, 1822]  Geinitz, L.: Über den Lehmbau im Altenburgischen. In: Monatsblatt für Bauwesen und Landesverschönerung, Heft 3/1822, Munich 1822

[DIN, 1947]  DIN 1169, 1947-06, Ausschuß für einheitliche technische Baubestimmungen (ETB)

## Chapter 11 : Building legislation and business practice

[Rath, 2004]  Rath, R.: Der Lehmbau in der SBZ/DDR, unpublished diploma dissertation at the Technischen Universität Berlin, Fakultät VII – Architektur Umwelt Gesellschaft, Berlin 2004

# Index

**abrasion** 16, 38, 42, 78, 82-83, 86-88, 153, 168

**absorption sealer** 48-49

**additives** 4, 13, 18-19, 23, 25, 27-31, 36-38, 40-42, 76, 81, 83, 87, 90, 95-97, 102, 106, 120, 152, 178, 196, 204-205, 223, 230, 240-241, 257

**adhesion strength** 17, 33-34, 82-83, 86, 88

**after-mixer unit** 66

**aggregates** 7, 14, 25-27, 29-32, 36, 40, 53, 88, 117, 121-122, 124, 134, 138, 142-143, 152-154, 158, 160, 162, 166, 168, 178, 190-191, 196, 250, 270

**air tightness** 77, 88-89, 115-116, 229

**alluvial soils** 8, 19

**ball test** 15, 18

**binding strength** 9-10, 13, 15-19

**brick wire mesh** 50, 63

**building material class** 33, 36, 88, 90, 106, 142, 149, 154, 171, 210, 240, 254

**bulk density** 27-30, 32-36, 81, 86, 88-90, 96-97, 103-104, 112, 114, 117, 120-121, 124, 128-129, 131, 138, 142-145, 149, 153, 171, 178, 191, 209-210, 239-241, 253

**capillary transport** 35, 112, 126, 201, 213

**cation exchange capacity** 11, 13

**ceiling infill** 100-103, 219, 242, 247-248

**ceiling insert** 246-247, 254

**cellulose** 27, 31, 35, 41-42, 91, 260, 272-273

**cement** 5, 17, 29, 31, 33-34, 38, 41, 44, 46, 55-56, 76, 80, 127-128, 138, 157-158, 161, 163, 168-169, 183, 188, 203, 206-207, 224, 233-234, 274

**chases** 46, 100, 125, 144, 147

**clay mineral analysis** 13

**clay minerals** 5, 7-9, 11-14, 16-17, 20-23, 28-29, 31, 33, 36, 42, 44, 60, 91, 95, 154, 201, 217

**coloured earth plaster** 51, 71, 78, 266-267, 272

**compaction** 120, 156, 158-159, 168-169, 176-177, 187, 190-191, 255

**compressive strength** 9-10, 17-19, 33-34, 78, 81-82, 86, 88, 104, 124, 138-139, 143, 146, 150, 153-154, 167-168, 191, 197, 201-202, 209, 233

**compressive strength class** 138-139, 150

**condensation** 9, 72-73, 75, 101, 109-112, 114-115, 120, 126, 130, 171, 186, 238, 249

**consistency** 15, 17, 19, 22, 24, 32, 41, 63, 67, 86, 91, 112, 125, 152, 156, 177-178, 216-217, 225, 243, 247-250, 270

**construction duration** 77, 160

**contact test** 140-141

**contaminant** 84

**corner bead** 68, 267, 272

**cost** 64, 72, 110, 151, 212, 226, 264-265, 272, 274

**cost calculation** 64, 72, 110-113, 148, 151, 212, 226, 264-265, 272, 274

**cow dung**  31, 205, 217

**crack formation**  26, 39-40, 42, 44-46, 50-51, 54, 60-61, 66, 69, 71, 75, 78-79, 81, 126, 140-143, 147, 152, 159-160, 163-165, 168-169, 189-190, 199, 206, 235, 237, 270

**cracks**  26, 33, 44, 47, 50-51, 55, 59-61, 73, 77-78, 89, 125, 141, 143, 160-161, 168-169, 185-186, 200, 205-207, 221, 237, 255

**crushing**  15, 23

**cutting test**  15

**damp proof course**  113, 147, 162, 168, 181, 191-192, 201-203, 208

**dimensioning**  34, 111, 146-147, 149, 167

**dip tests**  140

**drying**  10, 14, 26-29, 33, 44, 47-48, 60, 66, 71-77, 95, 100, 110, 112, 114-115, 119-125, 127, 143, 154, 156, 160-161, 168, 181-182, 213, 226-227, 245, 249-250, 255, 267

**drying protocol**  73-75

**drywall construction**  273

**earth blocks**  2-4, 21, 27-28, 31-34, 36, 45, 90, 95, 99, 101-106, 123-124, 133-147, 149-150, 154, 173, 177, 179, 184, 186, 191, 194-199, 202-205, 208-210, 214, 220, 225-231, 235-236, 239-240, 250, 263-264, 268-271, 273-274

**earth brush-on plaster**  42, 78, 91-93, 267, 272-274

**earth builder**  176

**earth building board**  3, 27, 45, 96, 100, 103-107, 145, 252, 254, 268, 273

**earth building codes**  3, 133

**earth loaf**  195

**earth masonry mortar**  30, 123-124, 129, 131, 133, 143, 196-197, 225, 268-271

**earth paint**  92, 267

**earth plaster**  5-6, 19, 23, 27, 30, 33, 36, 38-42, 44-48, 51, 53, 59, 62-63, 65-67, 69-73, 76-89, 91, 100, 116, 123, 125-126, 130, 141, 143, 205, 220, 225, 234-236, 240, 247, 254, 257-260, 266-267, 272-274

**earth plaster mortar**  30, 42, 51, 67, 71, 84, 116, 126, 225, 235, 267

**earth reel**  219, 242, 244-245, 253

**earth render**  205, 235

**edges**  51, 59, 62-63, 68-69, 110, 114, 144, 159, 164, 195, 204, 215, 221, 226, 229, 235-236, 243, 249, 267, 272

**erosion**  161-162, 182, 194, 203-205

**expanded metal**  50, 236

**extrusion method**  45, 96-98, 134, 138, 196-197, 229

**factory-made mortar**  41, 81-82, 85, 87

**figure-8-shape test**  17-18

**fixture**  69, 97, 100, 103, 118, 122, 124-127, 144, 192, 224, 238, 260, 265-269

**flatness tolerances**  52-53, 58

**floors**  116, 151, 168-170, 173, 180, 255-256

**formwork**  45, 117-121, 156-157, 160, 163, 175-176, 188, 249, 268-270

**frost proofing**  139, 154, 161

**frost resistance**  139, 140, 205

**geogrid**  159, 167, 169

**glacial soil**  8

**grain-size distribution**  16, 26

**grinding**  23

**grooves**  106-107, 182, 214-215, 218, 226, 228, 241-244, 271

**hammer scale**  255

**hand-made**  1, 134, 196-197

**hydrophobic treatment**  44, 161

**illite**  12

**in-situ soil testing**  14

**internal insulation**  2, 27, 48, 109-111, 113-115, 117, 121, 123, 125, 127, 129-131, 170, 209, 223, 268-269, 273

**kaolinite**  11-12, 16

lehmbau regeln   3-4, 18, 73, 80-82, 85, 87, 90, 112, 117, 121, 123, 126, 131, 133-135, 140, 143, 146, 151, 154-155, 160, 167, 171, 205, 227, 229, 241, 254, 262, 264

light earth   30, 32, 34, 36, 44-46, 61, 111-112, 117-124, 129-131, 138, 143, 264, 268-269, 271, 273

lime content   18-20

lime plaster   34-35, 59, 80, 166, 183, 198, 222, 235, 240, 251, 271-274

lime render   183, 198, 224-225, 231, 233, 235, 237, 272

lintel   118, 144, 149, 163, 185, 191-192, 269

loadbearing structure   144, 151, 153, 167, 179

loess soil   8, 16, 19, 23, 177, 190-191

loose earth fill material   252-253

measure of shrinkage   16, 19, 29-30, 33, 69, 81, 86, 143, 153-154, 160, 168, 190

mixed masonry   199

mixer   23, 32, 63-66, 152, 156

modulus of elasticity   33-34, 139, 153-154, 171, 179, 191, 209

moisture sorption   33, 36, 44, 85

moisture susceptibility   26, 29, 31, 38, 139-140, 154

montmorrilonite   12

mortar key   228

mould formation   28-29, 72-73, 75-77, 114-115, 121-122, 161

needling   206-207

ox blood   255

paddle mixer   63, 152, 156

paint   1, 31, 36, 38, 43, 47, 51-52, 55-57, 78-79, 81-83, 85, 91-93, 114, 134, 171, 205, 222, 232-234, 236, 238, 241, 259, 267, 271-274

panel infill   27, 110, 124, 130, 144, 178, 211-214, 219-221, 223, 225-227, 232, 237-239, 241-243, 254, 271, 273

perforated bricks   35, 46, 135, 137-138, 145, 229-230, 274

pest infestation   200, 208

plaster lath   45, 48-50, 90, 119-121, 126, 224, 232, 236-238, 243, 246-249, 252, 257, 259, 265-266, 268-269, 273

plaster reinforcement   47, 49, 59, 61-63, 77, 98, 126, 206, 237-238, 266-267

plaster system   20, 59, 79, 91, 183-184, 234, 238

plinth   147, 162, 176, 180-181, 191-192, 201

poisson's ratio   139

preparation   1, 4, 13-15, 21-24, 28-29, 45-46, 48, 51, 63-65, 71, 78, 92, 153, 155, 166, 169, 177-178, 206, 233-234, 237, 257, 259, 261, 266-267, 269, 271

press-moulding   134

primer   44, 46-49, 51, 55, 79-80, 92, 267, 272-274

quality classification   52-53, 77, 85, 91, 266-267

quality control   87

rammed earth   2, 6, 9-11, 17, 19, 22, 25-27, 29-34, 44-45, 134-135, 151-173, 175, 177, 187-195, 198, 200, 203, 207-210, 255, 263-264, 270, 274

rammer   134, 156, 158-159, 255, 270

reinforcement   27-28, 35, 40, 45, 47, 49, 57, 59-63, 69, 77, 95-98, 126, 143, 159-160, 164-165, 178, 205-207, 237-238, 266-268, 274

ring beam   147-148, 163-164, 185, 206

salt contamination   20-21, 43-45, 47, 114, 134, 201-202

scanning electron microscope   12

sealing   22, 109, 114, 116

shrinkage   1, 12, 14, 16-19, 26-27, 29-30, 33-34, 40-41, 44, 46-47, 54-55, 59-60, 66, 69, 71, 73, 77, 81, 86, 95, 143, 153-154, 160-161, 165-166, 168, 179, 181, 185, 188, 190, 195-197, 203-204, 206-207, 221, 229, 233, 255

**shuttering** 44-46, 118-121, 156-160, 166-167, 175, 177, 188, 190-191, 193-194, 214, 270
**sieving** 16-17, 22-24, 257
**silt** 16
**site-made mortar** 41, 63, 81, 83, 87
**slurrying** 23
**smectite** 12
**smell test** 15
**soaking** 22-23
**soiling** 67, 102, 232
**stabilised earth** 31, 41-42, 48, 54, 79-80, 82-84, 87, 133, 151, 163, 222
**stacked wall** 11, 98-100, 135, 264, 268, 273
**starch** 31, 41-42, 91
**staves** 27, 184, 214-216, 218-219, 223-224, 226-227, 239, 242-245, 248-250, 252-254, 271
**stiffening** 103, 144, 146-148, 179, 238, 252
**straw-clay** 90, 112, 117, 175-176, 178, 185, 214-219, 221-227, 235, 239-245, 247-250, 252-254, 257, 263, 271
**subconstruction** 48, 95-97, 103, 105, 118, 146, 252, 268, 273
**substrate** 41-51, 53-56, 59-61, 63, 70-71, 73, 75-77, 81, 83, 91-93, 100, 120, 125, 183, 204-206, 222-224, 231, 233-235, 237-238, 252, 257, 266-269, 271-272
**suction test** 140-142
**surface stabilisation** 78, 82-83, 87, 259
**surface treatment** 37, 50, 52, 57, 70, 77, 81, 85, 92, 98, 128, 152, 218, 232, 234, 258, 265-267, 270-272
**swelling** 10, 12-13, 45, 48-49, 92, 98, 135, 139-142, 164, 201, 203, 229, 233, 248-249
**tempering** 217
**tensile bending strength** 197
**thermal capacity** 33, 35, 104

**thermal conductivity** 27-29, 33-35, 111, 113-114, 119, 121, 128, 149, 170-172, 208, 210, 239
**three layer clay minerals** 9, 11-13, 17, 29, 33, 36
**tiles** 54, 80, 96-97, 114, 175, 193-194, 238
**timber-frame construction** 179-181, 185, 211-213, 215, 219-220, 226, 230, 232-233, 235, 237, 270, 273
**topcoat plaster** 42, 46, 51, 54, 57, 59, 68-69, 82-83, 166, 257, 259, 266-267, 273-274
**trial surfaces** 45-46, 77-79, 82, 155, 166
**triangular batten** 144, 157, 227-229, 270-271
**two layer clay minerals** 11-13
**undercoat plaster** 47, 54-56, 59-61, 68-69, 77, 87, 166, 183, 198, 223, 235, 266, 269
**usage class** 45, 98, 101-102, 104, 123, 134-142, 144, 205, 229-230
**wall heating** 43, 48, 62-63, 95-97, 111
**wallpaper** 55-57, 78-79, 91, 114, 267, 272-274
**wall replacement** 133, 202-203, 205, 224, 248
**wash test** 15
**water absorption coefficient** 33, 35, 236
**wattle** 214-219, 223-227, 239, 271
**weathering (summer/winter)** 22
**weller earth construction** 2, 27, 30, 33-34, 173-186, 190-192, 194-195, 197-198, 201-205, 208-210, 244
**work times** 265
**x-ray diffraction** 13